Biometric Data in Smart Cities

Sensors Communication for Urban Intelligence
Series Editors: Mohamed Elhoseny and Xiaohui Yuan

Biometric Data in Smart Cities
Methods and Models of Collective Behavior
Stepan Bilan, Mykola Bilan, Ruslan Motornyuk, and Serhii Yuzhakov

For more information about this series, please visit: https://www.routledge.com/our-products/book-series

Biometric Data in Smart Cities

Methods and Models of Collective Behavior

Stepan Bilan, Mykola Bilan, Ruslan Motornyuk, and Serhii Yuzhakov

CRC Press
Taylor & Francis Group
Boca Raton   London   New York

CRC Press is an imprint of the
Taylor & Francis Group, an **informa** business

First edition published 2022
by CRC Press
6000 Broken Sound Parkway NW, Suite 300, Boca Raton, FL 33487-2742

and by CRC Press
2 Park Square, Milton Park, Abingdon, Oxon, OX14 4RN

CRC Press is an imprint of Taylor & Francis Group, LLC

ISBN: 978-0-367-65024-7 (hbk)
ISBN: 978-0-367-65025-4 (pbk)
ISBN: 978-1-003-12746-8 (ebk)

Typeset in Palatino
by SPi Global, India

Contents

Preface

We live so close in this city ... and so far away...

In the conditions of an increase in the number and density of information flows that surround a person in the modern world, the processes of automation of their processing are of increasing importance. Today's world cannot be imagined without continuous human interaction with means of computational information processing, endowed with elements of artificial intelligence. In this regard, more and more aspects of life are becoming digital, which entails the need and importance of using of information security systems. Information security systems are implemented on many physical principles for the presentation and storage of information, and also rely on modern methods and means of computational processing and transmission of information. One of the main components of information security is biometric identification and personal authentication.

Biometric identification of a person is necessary to ensure authorized access to various types of data (personal, restricted, and secret). Methods and means of biometric data processing are rapidly being introduced into the daily life of people. The global biometric technology market is constantly growing, and according to a published report by MarketsandMarkets, it will continue to grow at approximately 13.4% between 2020 and 2025. At the same time, it is predicted that in 2025 the market for biometric technologies will double to about $ 68.6 billion. Almost every person on earth constantly encounters and interacts with biometric technologies using smartphones and other automated means of human life, using contact and contactless means of biometric identification.

One of the characteristics of modern society is high urbanization and an increase in the urban population and the population of the entire globe. This natural phenomenon also leads to the indisputable need for the use of biometric technologies. In addition, the growth of the urban population has provoked a new direction that determines the creation of smart cities aimed at improving the comfort of human life in the urban environment and automating many life processes.

Smart city technologies use distributed intelligent systems that rely on a network of smart sensors, IoT, and a high degree of data processing. Now this is a relatively young direction, and for its development as well as for the implementation of various kinds of projects, the search and development of new methods and means for their implementation is required. Biometric technologies take one of the main places in smart city technology.

This book aims to describe biometric technologies that are used in the smart city environment. The biometric identification methods described in the book use the theoretical foundations of the parallel shift technology, the Radon transformation implemented on hexagonal cellular automata, and the theory of cellular automata with active cells.

Parallel shift technology is a way of researching information, which was proposed by the authors at the beginning of the 2000s. This technology can be used in pre-processing and video analysis. In combination with digital processing of images, it can enhance system performance machine vision.

The book discusses certain methods of applying parallel shift technology, which are a continuation of previous research on this topic. One of the important components of information processing is its reproduction from the saved data or transformation into the form more acceptable for the further analysis. In this direction, the authors are conducting research. In addition to information protection systems, this technology can be used in robotic visual systems, automatic transport control systems, and in medicine for eye prosthetics.

Radon transformation technology, implemented on cellular automata with a hexagonal coating, has found its effective application in biometric personality identification systems, as well as in modeling dynamic processes and collective behavior of various biological colonies. Cellular automata with active cells are also effectively used to model the dynamics of the behavior of colonies.

The book covers the basics of building distributed intelligent systems and multisensor systems. Sensors that are used in a smart city environment are described, and the foundations of building a unified urban biometric community are considered, and models of the behavior of colonies of biological organisms based on cellular automata with active cells are presented.

The book consists of 13 chapters.

Chapter 1 describes the basics of building distributed intelligent systems and also describes the characteristics of the organization of various biological colonies that exist in nature and their behavior.

Chapter 2 describes the general provisions of the interaction of the system with the external environment, provides a description of multisensor systems, and gives a broad classification of sensors according to various criteria.

Chapter 3 is devoted to an examination of the main characteristics of a smart city, as well as the tasks that are solved in a modern smart city.

The substantiation of the main types of biometric sensors is given and their role in improving the comfort of residents and guests is emphasized.

Chapter 4 discusses the main biometric characteristics that are used for biometric identification of a person.

Chapter 5 describes the principles of building a unified multisensory system in a smart city based on biometric sensors and also describes the principles of building a unified city biometric community.

Chapter 6 describes the main theoretical provisions of the parallel shift technology, and also describes some options for transforming visual information for further analysis. Particular attention is paid to the process of controlling noise in the image.

Chapter 7 describes the processes of restoring the original image using the data of the reference surface using the circumscribed rectangle method. Methods for determining the parameters of the described rectangles for figures of various shapes are considered.

Chapter 8 discusses the processes of determining the main parameters of the scene using the parameters of the cyclic functions of the area of intersection. Ways to optimize the process of determining the shape of scene objects by dividing this process into two stages are considered.

Chapter 9 describes the main theoretical provisions that are used for biometric identification of a person in a smart city. The mathematical apparatus of the Radon transformation implemented on a cellular automaton with a hexagonal covering is described, and the main advantages are presented.

Chapter 10 describes methods of biometric identification based on images of the geometric shape of the auricle. Described are two options for biometric identification.

Chapter 11 presents methods of biometric identification of a person by the dynamics of his gait. To do this, in each frame of the video sequence, the human silhouette is processed using the parallel shift technology.

Chapter 12 presents a method for organizing a digital network of a smart city based on the pulse-time method for describing complex images. It describes a network consisting of biometric sensors located on the territory of a smart city divided into sectors. The formation of a general urban picture is described, which allows you to control the movement of selected objects in real time.

Chapter 13 describes the theoretical provisions of the technologies of cellular automata with active cells. Models of behavior of colonies of living organisms are presented. Models of formation, movement and destruction of colonies, as well as models of interaction (union and destruction) of colonies with different active cells are considered.

This book is intended for undergraduate, graduate students, and specialists working and conducting research in the field of biometric information processing, as well as in the development and construction of distributed intelligent systems. The book will be useful for all people living in modern smart cities and guests, as it will help them better understand life in a smart city, and can also serve them to improve comfort in the development of the city itself.

Stepan Bilan
State University of Infrastructure and Technology, Ukraine

Mykola Bilan
The Municipal Educational Institution Mayakskaya
Secondary School, Moldova

Ruslan Motornyuk
PU "Kiev Department" Branch of the Main Information and Computing
Center of the JSC "Ukrzaliznytsya", Ukraine

Serhii Yuzhakov
Main Department of STS in Vinnytsia Region, Ukraine

About the Authors

Stepan Bilan was born on September 15, 1962 in Kazatin City, Vinnytsia Region, Ukraine. He studied at the Vinnytsia Polytechnic Institute from 1979 to 1984. In 1984, he graduated from Vinnytsia Polytechnic with honors and an engineering diploma specializing in Electronic computing machines. From 1986 to 1989, he studied at the graduate school of the Vinnytsia Polytechnic Institute focusing on computers, complexes and networks. In 1990, he defended his thesis by specialties: 05.13.13 – computer systems, complexes and networks; 05.13.05 – Elements and devices of computer facilities and control systems. He worked at Vinnytsia State Technical University (now the Vinnytsia National Technical University) from 1991 to 2003. In 1998, he was awarded the academic title of assistant professor of "Computer Science" at Vinnytsia National Technical University. Since 2003, he has worked at the State University of Infrastructure and Technology (Kiev, Ukraine).

Mykola Bilan was born on June 27, 1961 in Kazatin, Vinnytsia Region, Ukraine. In 1983, he graduated with honors from the Vinnytsia Polytechnic Institute and has a specialty in Automation and Remote Control. Until 1994, he worked as a production master in the final assembly shop at the Tashkent Aviation Production Association, V.P. Chkalov. In 1989, he was the best master of the Association. From 1994 to 2015, he worked at the Radio and Television Center (Republic of Moldova) as an engineer and the head of the dispatch service. Currently, he works as an informatics and physics teacher in a secondary school in the village of Mayak, Republic of Moldova. He has many inventions, publications, and patents in the field of cryptography, steganography, and computer technology.

Ruslan Motornyuk was born on June 20, 1976 in the village of Ivankovtsy, Kazatinsky District, Vinnytsia Region, Ukrainian SSR. In 2001, he graduated from the magistracy of Vinnytsia State Technical University and received a master's degree in computer systems and networks. In 2013, he defended his thesis for the degree of candidate of technical sciences in the specialty "Computer systems and components". Currently, he works as a leading engineer in the Production Unit "Kiev Department" branch of the Main Information and Computing Center of the JSC "Ukrzaliznytsya".

Serhii Yuzhakov studied computer engineering at Vinnytsia State Technical University from 1992 to 1997.

Since 2002, he has been working in the information technology divisions of various state institutions of Ukraine. He has also been engaged in research in the field of pattern recognition.

In 2014, he received a Ph.D. with a specialization in "Computer systems and components".

1

Distributed Intelligent Systems and Natural Collective Intelligent Systems

1.1 Introduction: Background and Driving Forces

Distributed intelligent systems (DISs) is a new direction in artificial intelligence (Bedrouni, Mittu, Boukhtouta, & Berger, 2009; Lesser & Ortiz, 2003; Müller & Dieng, 2012; Kulkarni, Tai, & Abraham, 2015). The development of computer technology led to the creation of intelligent systems, which in their evolution became more and more complex and more and more difficult they solved problems. This led to the creation of DISs. It is believed that this direction arose in the 80s of the last century.

Distributed problem solving systems are based on the decomposition of the complete problem into partial problems. Each partial problem is handled by a standalone local solver. Through the interaction of local solvers, the obtained local solutions are combined, which makes it possible to form a solution to the complete problem. This distributed process makes it possible to solve much more complex problems than a single centralized solver can do.

Modern information technologies and communications have led to the need to use DISs that allow making decisions in an environment where many parts of the system are located at different distances, but are a necessary part of an intelligent system. Also, all elements of the system can be located and combined in one separate module, while each solves its own problem. An example would be an image recognition system, which is an array of processing elements. If you remove such a processing element, the problem may not be solved correctly. The practical complexity of modern computing systems and systems for the interaction of robots, intelligent agents (IAs), and other elements makes us talk about distributed computing and distributed artificial intelligence.

1.2 DIS Short Classification

Currently, DIS is classified according to many characteristics. However, in general, DIS can be divided into three types:

- systems of distributed solving of problem;
- distributed systems of problem solving;
- DIS with a variable number of IAs.

The first class of systems is characterized by the fact that all components of the system solve finite individual problems, the results of which determine the solution of the problem. The results of solving these problems determine the solution to the problem of the entire DIS. In this case, each subtask is autonomous and is solved by autonomous elements of the system. In fact, the general solution to the problem is based on the results of solving sub-problems. Basically, such systems make a decision that is explicitly presented as a numerical value. As a rule, in modern literature, individual DIS modules are IAs that can be implemented both in software and hardware (Kulkarni, Tai, & Abraham, 2015; Boissier, Bordini, Hubner, & Ricci, 2020; Ryzko, 2020).

IAs can perform various functions from single-bit functions to complex intellectual functions. If all agents of the system perform predetermined calculations, the final result is also obtained from the results of the calculations without any data analysis. If in the final result there is no decision-making, but only a computational result, then the system is a computational distributed system. DIS solves an intellectual problem. As a result of DIS functioning, there is decision-making based on data mining.

Autonomous modules (agents) can solve both intellectual tasks and ordinary computational tasks. Such a system can solve many intellectual tasks by certain IAs. For different tasks, IAs have their own classifications, which depend on the functions performed. Accordingly, IAs have different properties. IAs can have individual and collective properties.

If all IAs of DISs have only individual properties, then such a system is limited and cannot solve the problems of evolution and collective behavior. Individual properties include autonomy, social behavior, reactivity, basic knowledge, beliefs, desires, goals, intentions, benevolence, truthfulness, and rationality (Woldridge & Jennings, 1994).

Also, IAs can have collective properties, which display the interaction of the IAs in the system. These properties include connectivity, coordination, cooperation, collaboration, and union formation (Kulkarni, Tai, & Abraham, 2015; Boissier, Bordini, Hubner, & Ricci, 2020; Ryzko, 2020).

The second class of DIS is characterized by the fact that distributed modules (agents) do not solve individual intellectual tasks, but operate in a certain mode and interact with each other. At the same time, they solve a common collective problem, which is a collective solution for the entire team of agents of the system. In such systems, the result can be presented implicitly and can form the further behavior of the system as a whole. As a rule, such systems solve one or more intelligent collective tasks that are acceptable for all DIS users.

The third class of DIS includes systems in which, under certain conditions, new IAs appear (are born), as well as existing IAs disappear (die). New IAs can have properties that are inherent in already existing agents in the environment or they can have new properties that none of the old IAs in the environment had yet. The set of properties can be determined by the DIS structure and intellectual tasks that the system must solve. Also, the environment in which the DIS operates can form new agents with new unpredictable properties. This environment is of interest to researchers. On the basis of such an environment, it is possible to simulate various dynamic processes in time and describe the evolution of the behavior of both one agent and the whole team.

As mentioned earlier, DIS solves intellectual problems. Therefore, it is important to identify those intellectual problems for which the use of DIS is the most effective approach. First of all, such tasks include tasks of collective behavior that are inherent in the natural living organisms of their colonies. Modern animals constantly solve complex problems aimed at survival (searching for food, identifying danger, and other tasks). However, the most difficult tasks are solved by humans and communities of people.

In the process of life, a person constantly solves complex intellectual problems, and also creates new tools that help a person make decisions for further behavior. However, people always form collective communities (family, city, state, etc.) and therefore people are forced to solve the problems of the entire team. These tasks require constant interaction of people and other functions. Without such collective interaction, the collective task cannot be solved, which can lead to the death of the entire colony.

Most of the collective tasks cannot be solved without the use of new IAs, which have new properties and can also gain experience in solving problems from the IAs that created it. The solution of such problems is based on flexible DIS, which can change their structure depending on the appearance and disappearance of IAs.

In general, all existing collective biological systems combine all three types of DIS, which make up a complex intelligent system. The most developed systems of distributed intelligence make up associations of people who solve global problems for comfortable survival and prolonging the life of both an individual and the entire community. The study of the behavior of such systems on the basis of various models will make it possible to create more perfect associations that combine interacting biological objects and artificially created DISs, which partially replace the functions of individual biological objects and also help them make decisions in future behavior.

1.3 Natural Collective Intelligent Systems

All modern collective intellectual systems that exist in nature are characterized by a different form of organization. The most common form of organization is the selection of a leader and the subordination of all members of the biological colony to him or her. The leader develops a strategy of behavior and in most situations independently makes decisions about the further behavior of the entire colony. There are also colonies in which a decision is made by a majority vote. One of the forms of colony organization is the interaction of its members on the basis of mutual concessions, which make it possible to take a modified intermediate decision. There are colonies that maintain their resilience based on consensus. All members or leaders of the colony make a decision only if they are all for this decision and there is not a single vote against it. And one of the most highly organized forms of the colony is a cooperative operation based on the principle of "winning together".

An important factor in the colony is the survival factor. There are colonies without which an individual dies, and there are colonies that consist of individuals that can live alone. Colonies are created for different purposes: for reproduction, protection, food, etc.

There are many colonies that are homogeneous. For example, colonies of different ants. However, they never unite, but can destroy and enslave each other. There are also colonies of different individuals that get different prey (food) in different ways.

Colonies of different individuals lead different lifestyles. Moreover, some colonies can feed on individuals of another colony. The task of the predator colony is to find weak colonies and eat their members. The task of non-predator colonies is to protect all members of the colony and implement collective methods of protection. For example, a school of fish forms the shape of a colony and a trajectory of movement that protects against attacks by predators. The predator colony also solves the problem of distributing colony members when capturing non-predator colony members. In this situation, we are talking about proportionate colonies.

1.4 Ant Colony

One of the most studied natural colonies is the ant colony (Sun, 2011). Ant colonies investigated find the shortest route to a food source. The essence of the behavior is that the ant, which is the first to find food,

FIGURE 1.1
An example of a collective movement of ants. (A fragment of a video taken from the site https://www.youtube.com/watch?v=-NEAVhc9Eno)

returns home and leaves behind pheromones. Over time, pheromones evaporate. On the shortest path, pheromones last longer. Pheromones attract nearby ants, which fix the designated path by depositing their pheromones each time. Long paths eventually disappear and only the shortest paths remain. The exchange of information in the ant colony is carried out through pheromones. An example of an ant's movement is shown in Figure 1.1.

Thus, ants with weak intelligence solve the difficult problem of survival through distributed collective interaction. In addition, in an ant colony, various responsibilities are distributed between ants. Each ant individually performs its duties (scout, worker ant, soldiers, nannies, pastoralists, builders, foragers, etc.).

The responsibilities that a particular individual is endowed with depend on its individual qualities. More initiative and mobile ants become scouts and hunters. Their reactions help provide food for the community. Calm and sedate representatives of the family are engaged in grazing aphids and collecting its syrup. The age of the insect can affect the change in its activity.

The common goal for all individuals is the life support of the anthill. The well-coordinated work of all ants and a clear distribution of their social roles and tasks that they solve allow ant colonies to survive and reproduce. This distribution of roles makes the ant colony as a whole a highly organized distributed intellectual system of nature.

1.5 Bee Colony

The bee colony solves problems collectively. The colony itself is built in such a way that the entire colony is a single mechanism. Each bee cannot survive alone. The goal of the entire bee colony is to find and collect nectar in the area adjacent to the hive. There are different types of bees in the colony, which have different functional loads. A group of bees conduct preliminary exploration of the area adjacent to the hive for the presence of a large amount of nectar. These bees, through dance, inform the worker bees of the location of the places where the largest amount of nectar was found. Also, certain functions are performed by drones and the uterus.

The algorithm of the bee colony is to determine the most promising points with the best values of the objective function (Karaboga, 2005). In the adjacent territory of certain points, exploration is carried out. The information is updated in accordance with the received data. After the end of the entire operation of the algorithm, the search for the best solutions is carried out.

1.6 Collective Movements in Nature

Figure 1.2 shows examples of collective movements of animals and people in nature.

This figure shows examples of the coordinated movement of a group of animals and people. Almost all moving groups observe a geometric shape and can collectively change it depending on the situation. Individuals of different ages and sizes are involved in these groups. In addition, the groups consist of a different number of members. To solve this problem, each member of the group takes his or her own place and performs a specific function.

In such movements, there are no collisions between the members of the pack, and the packs quickly react to obstacles and go around them.

Much attention is paid to the movements and behavior of colonies, whose members have intelligence. The goals and objectives of such colonies are different. However, they take into account the state of the external environment as well as the presence of other intellectual colonies.

In addition, individuals with intelligence also consist of small living organisms, which together form intelligence. For example, the human brain consists of billions of neurons and connections between them. It is believed that one neuron processes one bit of information, and the whole brain solves very complex problems. This feature will not be studied soon. However, there is no doubt that the brain is a distributed biological intellectual system that makes it possible to control human behavior, as well as the behavior of a

FIGURE 1.2
Examples of collective movements in nature. (Images taken from the site www.Bing.com/images/)

whole group of people. The interaction of human intelligences allowed in the process of evolution to create a biological highly intelligent system, which is represented by human society or a part of it. People create intelligent environments that shape new intellectual approaches for people to interact. One of such environments is a smart city, in which people are rebuilt and their behavior differs from the behavior of people in cities that do not possess the properties of urban intelligence. How human intelligence is rebuilt depending on the increase in the intellectual properties of the environment is still not fully known. However, it can be confidently asserted that the behavior of people in a smart environment differs significantly from low-tech environments that do not use the full digitalization of all spheres of human activity.

2

Multisensor Systems

2.1 Interaction of the System with the External Environment

Everything in the world is interconnected. Let's consider an abstract system (Figure 2.1). Any system "lives" in the surrounding space, has a goal, has or itself develops an algorithm for achieving the goal, and interacts with the surrounding world.

The goal for any system can be to change any parameters of the system itself or the environment. By and large, the goal is set by the person. But the system itself can set the goal. Such systems should have elements of artificial intelligence. The communication between the system and the target is duplex, i.e. bidirectional. During the operation of the system, the goal can be adjusted.

The algorithm for achieving the goal in many cases is also formed by a person. Systems with elements of artificial intelligence can form an algorithm for achieving the goal as well as adjust it in the process. Consequently, the connection between the goal achievement algorithm and the system is the same duplex.

And now the most important thing is the interaction of the system with the external environment. The external environment is characterized by a large variety of parameters and, to a greater or lesser extent, affects the operation of any system, regardless of its nature and goals. The parameters of the external environment are characterized by physical quantities. An example of physical quantities that affect the operation of the system can be temperature, time, pressure, distance, illumination, noise level, force, speed, and tens or even hundreds of other quantities. In Figure 2.1, the influence of the external environment on the system is shown by many arrows.

The system can also influence the external environment to achieve the goal. But this effect is small (in Figure 2.1, it is shown with just two arrows) compared to the reverse. For example, if a person needs to go through the jungle, then he can go around them or, armed with a machete, go through the "impenetrable" jungle. Each of these decisions is related to the use of system resources, time, and other factors. It is not for nothing that the people say:

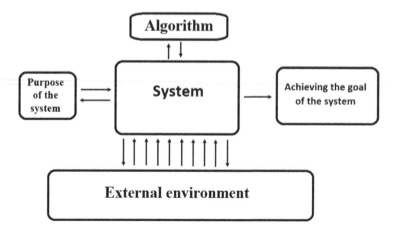

FIGURE 2.1
The structure of an abstract system.

"The clever will not go up the hill, the clever will bypass the mountain" or "If you hurry, you will make people laugh". The irrational use of resources can lead not to the achievement of the goal, but the irrationality of the use of time, to its untimely implementation.

So, we found out the fact that the external environment significantly affects the operation of the system, and, therefore, largely determines the possibility of achieving the goal. The question arises as to how to assess the impact of the external environment on the system? Trying to get an answer to this question, we come to the conclusion that it is necessary to somehow measure and evaluate the parameters of the external environment. To do this, you need to use some devices that allow you to do this. And such devices do exist, and they are called sensors. Let's try to give a clear definition of sensors.

A sensor is a device for measuring any physical quantity or its change and converting this quantity or its transformation into a signal for use by the system when reaching a goal. In most cases, such a signal is an electrical signal.

2.2 Sensor Classification

Since the external environment affects the system with a large number of physical quantities, there are a large number of sensors. Let's try to classify sensors according to different criteria.

I. According to the shape of the output signal, the sensors are:
   a) analog in which the output signal is an analog or continuous form;
   b) digital, the result of work, which is a digital signal.

II. By the type of measured physical quantity, sensors can be classified as follows:

a) temperature sensors (temperature sensors);

b) pressure sensors;

c) speed sensors;

d) time sensors (timers);

e) steering angle sensors;

f) light sensors;

g) sound sensors;

h) distance (height) sensors;

i) touch sensors (tactile sensor);

j) color sensors.

Each of the indicated groups of sensors includes several varieties, depending on the principle of operation of the sensitive element and design.

Let's consider some of them. So touch sensors or tactile sensors can be made using mechanical contact closure, one of which is made in the form of a soft spring, the second in the form of a rigid rod. When touched, the spring bends and closes to a hard contact, which is interpreted as a touch.

Force sensors are another example of tactile sensors. The operation of such sensors is based on the piezoelectric effect, which involves the fact that when a force is applied to a piezoelectric material, a potential difference appears on its opposite surfaces. The greater the applied force, the greater the potential difference. Such sensors record not only the fact of touch, but can also assess the force that is applied to the device from the environment.

Distance sensors are used to measure distances, which, depending on the type of operating signal, are divided into infrared (IR range finder), laser (laser range finder), and ultrasonic (ultrasonic range finder). The principle of operation of ultrasonic sensors is based on the phenomenon of echolocation. Such sensors are equipped with an ultrasonic emitter and a receiver. The emitter emits an ultrasonic pulse toward the obstacle, the timer measures the time until it returns to the receiver and, knowing the speed of sound propagation in the environment, the distance to the obstacle is calculated. The disadvantage of such sensors is the large angle of the directional pattern, which leads to errors in measuring the distances to obstacles with irregularities. This disadvantage is absent in optical rangefinders in the infrared range. However, optical rangefinders, unlike ultrasonic ones, have difficulties in determining the distance to transparent and flashing objects.

Long distances can be measured with a laser rangefinder that works on the same principle as ultrasonic ones, with the only difference that instead of an ultrasonic pulse, they use a narrow laser beam.

FIGURE 2.2
Rotation angle sensors appearance.

A wide range of sensor classes are angle sensors. By the type of output signal, they are divided into analog and digital. Analog angle sensors are based on a potentiometer, which is a voltage divider consisting of a variable resistor with three leads, one of which moves on a resistive plate, thereby changing the resistance between the movable and fixed contact. The movable contact is connected to the shaft, the rotation of which changes its position and, therefore, the resistance (Figure 2.2).

The advantages of such sensors include the ability to measure the absolute angle of rotation. The disadvantage is the fact that over time the resistive plate oxidizes and wears out, which leads to the appearance of the so-called "dead" zones, in which the contact can disappear altogether or there is increased instability of the sensor reading.

Analog rotary angle sensors can be based on selsins, rotary transformers, magnesins, or inductosins.

Digital rotary encoders are often referred to as encoders. They are of two types: incremental and absolute.

An incremental encoder is a thin disc with alternating transparent and darkened areas applied to it. On one side of the disk is a light source (narrow laser), on the other – a photodetector. When the disc is turned, a pulse sequence is generated at the output of the photodetector, according to which the controller can determine the speed of rotation of the disc and the angle of rotation. The smaller the size of the transparent and dark areas, the higher the resolution of the sensor.

To determine the direction of rotation, a second photodetector is built into the sensor in such a way that if, at the start of movement, the first photodetector first opens, and then the second, then this rotation is in one direction, and if the second photodetector first opens, and then the first, then this movement is in the other direction. Such sensors are called quadrature sensors. They double the positioning accuracy. The disadvantage of incremental encoders is the fact that they require additional initialization of the initial position.

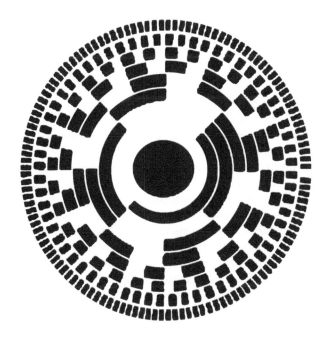

FIGURE 2.3
Absolute encoder disk. (Picture taken from the site https://indicator.ru)

The indicated drawback is eliminated in the absolute encoder. In such an encoder, the disk consists of tracks on which darkened areas are applied according to the following scheme. Each track, starting, for example, from the center of the disc, is divided into 2n sections, where n is the track number. The first track is divided into two sections, the second – into four, the third – into 8 sections, etc. Every second section on every track is darkened (see Figure 2.3). A photo detector is installed for each track. Therefore, at the output, we get a unique binary code for each position of the disc.

In addition to the classical binary code, the so-called Gray code is often used, in which each subsequent state differs from the previous one by only one bit. As a result, due to the ambiguity of the reading, the error is equal to the unit of the lower digit. Gray code – is non-positional, those the weight of the logical "1" is not determined by the digit number.

Sensors for determining the position in space deserve special attention. Such sensors include classical gyroscopes, gyroscopes using micro-electrical and micro-mechanical technologies (the so-called MEMS technology), and accelerometers. Let's consider some of them.

The principle of operation of the classical mechanical gyroscope is based on the law of conservation of torque. If you place a rotating body in a closed system, then the direction of the axis of rotation of the body remains unchanged.

In gyroscopes, the role of a closed system is played by a gimbal with the ability to rotate around all three axes, in which a rotating body is placed. Such a body is usually a massive disk rotating at high speed. Since there are no ideal closed systems, the greater the mass of the disk and the speed of its rotation, the smaller the deviation of the axis of rotation of the disk from the given direction. Therefore, for example, several gyroscopes oriented in different directions are installed in spacecraft. Onboard electronics compares the data from each sensor (gyroscope) and by averaging possible deviations accurately determines the orientation of the spacecraft in space. Classic gyroscopes are mainly used in aviation, submarines, spacecraft, and other systems where accurate position determination in space is required.

With the development of microelectronics, it became necessary to minimize position sensors in space, both in overall dimensions and in weight. There was a technology called the micro-electro-mechanical systems (MEMS) technology or microelectromechanical systems. MEMS technology is a technology that allows you to miniaturize mechanical structures and integrate them with microelectronic circuits into a single physical device, which is, in fact, a system (Iannacci, 2017). The main and dominant field of application of MEMS technology is sensors.

MEMS technology has made it possible to create a miniature gyroscope that is found in a huge number of smartphones, tablets, and other microelectronic devices. The MEMS gyroscope is actually a so-called vibrating pendulum, which, when the gyroscope turns, tries to resist. The resistance of the pendulum to the applied force is fixed and converted into an electrical signal. Sometimes the analog output from such a gyroscope is converted to digital using an analog – digital converter (ADC). MEMS gyros are often referred to as steering angle speed or gyrotachometers.

Another device that has emerged thanks to MEMS technology is the accelerometer, a device that measures the acceleration of a body under the influence of external forces (Tinder, 2007). Schematically, the accelerometer can be depicted as a relatively massive body freely moving along some axis (Figure 2.4).

The body is connected to the hull by springs. When the device is displaced to one side, the weight will move along the axis in the opposite direction. The magnitude of the effective acceleration is determined by the displacement of the body relative to the center of the axis. To measure a position in three-dimensional space, it is necessary to measure the projection of acceleration along three axes at once. Knowing these values, it is possible to calculate the angle of inclination of the device relative to the ground. However, accelerometers have one major drawback. The possibility of accurate measurements is determined by the inertiality of the system, i.e. provided that the accelerometer, and therefore the device, is at rest or moves uniformly and in a straight line. If an external force acts on the accelerometer during measurements, the device fixes it and thereby introduces an error in the readings of the tilt angle.

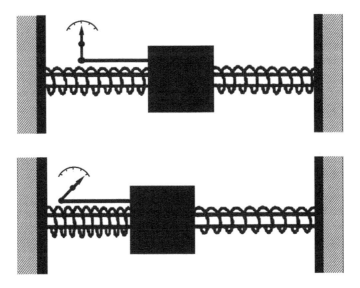

FIGURE 2.4
Accelerometer. (Picture taken from the site https://indicator.ru)

A few words about pressure sensors. Earlier, a force sensor was mentioned, which in some cases can be used as a pressure sensor. However, there are many pressure sensors that differ in the used sensing element, and, therefore, have a different principle of operation. Since there is a need to measure three types of pressure, the pressure sensors are divided by the type of pressure being measured into three groups:

1. Sensors measuring absolute pressure, which includes atmospheric and gauge pressure.
2. Excess pressure, which is the difference between absolute and atmospheric pressure.
3. Differential pressure, which is the pressure difference between two points.

The most accurate and sensitive are the fiber-optic pressure sensors (Yin & Yu, 2002). The principle of operation of their sensing element is based on a change in the refractive index of light when passing through a two-layer fiber-optic element (Figure 2.5a) with a variable gap (Figure 2.5b).

Currently, for measuring small displacements, promising are optoelectronic sensors that use the phenomenon of light interference, which consists in the addition of light waves. Such sensors, in addition to high accuracy and sensitivity, have ultra-small dimensions due to the small wavelength of light. Therefore, they are widely used in medicine.

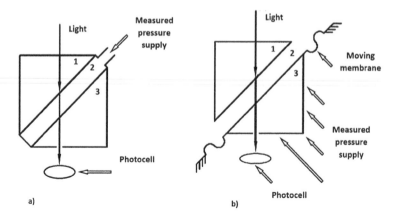

FIGURE 2.5
Fiber-optic pressure sensors. (Picture taken from the site https://electrosam.ru)

In addition to fiber-optic pressure sensors, capacitive sensors are widely used, the principle of operation of which is based on measuring the capacitance between the stationary and dynamic plates of the capacitor. Magnetic or inductive pressure sensors are also used, based on the measurement of magnetic flux between a stationary ferromagnetic core and the same movable membrane. Considering that the magnetic permeability of a ferromagnet is hundreds and even thousands of times higher than the magnetic permeability of air, a slight change in the gap between the core and the membrane leads to a significant change in the magnetic flux.

Depending on the characteristics of the substance, environmental conditions, pressure range, sensitivity, and accuracy, piezoresonance, mercury, potentiometric, and other pressure sensors are used, but we will not dwell on them.

Let's consider another class of sensors – temperature sensors. There are also a huge number of them, but mainly there are several main types. The first is a semiconductor crystal, the resistance of which is highly temperature-dependent. Such sensors are called thermistors (thermistor). They, in turn, are divided into low-temperature and high-temperature. The second class is called thermoelectric sensors or thermocouples (thermocouple). They are a compound of two metals with different temperature coefficients of resistance (TCR), which, in turn, characterizes the dependence of the relative resistance of the metal on temperature. At the junction of such metals, with different TCR, when heated, a potential difference appears. By amplifying this signal with, for example, an operational amplifier, you can use it to measure temperature. The next type of thermal sensors is the diode or, as they are often called, semiconductor. The sensing element in such sensors is a silicon diode. Under the influence of temperature, the resistance of the p–n junction changes and, therefore, the forward voltage, which is used to measure the temperature. An interesting type of temperature sensors are acoustic sensors. Their principle of operation is based on measuring the speed of sound in a medium when the temperature changes. Such sensors are non-contact, which allows them to be used in hard-to-reach places, high-risk objects, for example, in places with high radiation, as well as in medicine when measuring temperature, which is especially important now, in the context of the coronavirus pandemic (COVID-19).

Various types of sensors can be described for a long time. However, the examples given earlier are sufficient for realizing the fact that sensors represent a huge class of various devices.

2.3 Multisensory Systems

The modern rapid development of science and technology has led to more capacious, stringent, and complex requirements for the assessment of environmental parameters. At the same time, increased requirements are imposed on the accuracy of measurements, the speed of processing incoming information and, consequently, on the speed of decision-making. The use of individual sensors in technical systems has ceased to meet the demands of the modern level of human development. To solve problems at a higher level, the idea arose out of combining two or more sensors into a single system. Such a combination is performed not only at the level of obtaining information, but also at the levels of merging and data processing, as well as assessing environmental parameters and making decisions.

Systems that comprehensively assess the parameters of the environment or an object consist of many sensors, provide data fusion, integrate processing of incoming information, and are capable of evaluating and filtering information are called multisensor systems.

Over the past few decades, in connection with the rapid development of computer technology, which allows the processing of multiparametric information in real time, its cost-efficient, multisensor systems have received a sharp development.

Let's consider some of them. Until now, the organs of chemical senses (smell, taste), in contrast to the organs of physical senses (sight, hearing, tactile perception), have not been fully investigated and were surrounded by a halo of mystery. One of the representatives of multisensory systems that detect odor is the "electronic nose" (Patel, 2013). Examples of such sensors are shown in Figure 2.6.

"Electronic nose" is a multisensor system designed for quick and high-quality assessment of odors in real conditions. The main components of the

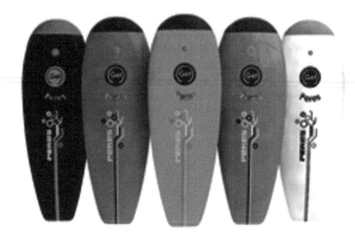

FIGURE 2.6
Electronic nose system. (Picture taken from the site https://www.popmech.ru)

"electronic nose" are a set of sensors that provide information on the composition and concentration of individual components in multicomponent mixtures. There are currently two main approaches. The first is to use sensors that only respond to a specific substance or compound in a mixture. The second approach is to use relatively non-selective sensors that respond to groups of similar molecules, mimicking the work of the brain. It is assumed that the different responses of the sensors, even if not entirely accurate, in the aggregate and after appropriate processing, will make it possible to estimate the content of the mixture. In addition, sensors that work on different principles are often combined in one system, expanding the capabilities of the device.

The fields of application of multisensor systems such as "electronic nose" are truly endless. "Electronic nose" can be used and is already being used in pharmacology to ensure quality control over the course of the technological process during the production and storage of drugs and in medicine for the express diagnosis of acute infections by the quality of exhaled air. In forensic science, it is used for the investigation of cases related to crimes against the person, in particular, in the determination of the sexual and individual characteristics of a person's odor by analyzing the various odors, e.g. sweat, blood, and hair. In the military field, it is used for the detection of toxic and biological agents. The "electronic nose" can be used by the security services of airports and customs terminals, in particular, to detect explosives and narcotic substances in passengers, prohibited from smuggling drugs, etc. during control checks.

An equally interesting representative of multisensory systems is the "electronic tongue" (Toko, 2003; Savoy et al., 1998; Riul, Dantas, Miyazaki, & Oliveira, 2010). Currently, there are many varieties of such systems. The main component of the "electronic language" is an array of sensors. Usually it includes 5 to 30 sensors made of a wide range of membrane materials such as plasticized polymers containing active substances and chalcogenide glasses with impurities of various metals. In addition, potentiometric sensors can be used in the "electronic language", as well as optical sensors, which are grains of polyethylene glycol–polystyrene rubber, derivatized with indicator molecules, etc. Chemical sensors with relatively low selectivity are used as sensors, i.e. having sensitivity to several components of the analyte simultaneously. These sensors are sometimes referred to as cross-sensitivity sensors.

A special place in multisensor systems such as the "electronic nose" and "electronic tongue", and others is occupied by methods of processing data from an array of sensors. Data from cross-sensitive sensors contain information about various components present in the environment. To extract this information, you need to have methods that allow you to analyze the responses of all sensors in the system together. Such methods can be methods of pattern recognition and multivariate calibration. The choice of a particular method depends on the task of the multisensor system. Typical data processing tasks are:

- problems of recognition and study of the data structure;
- problems of identification and classification of components of the environment;
- determination of concentrations and other quantitative parameters.

To solve the first problem, unsupervised learning methods are used, for example, principal component analysis and some types of artificial neural networks. Identification and quantitative analysis tasks are solved using guided learning methods. To determine the concentrations of components, calibration dependences are found using multivariate calibration methods, the most common of which include correlation by principal components and fractional least squares, as well as artificial neural networks.

The ability of electronic tongue systems to recognize and identify multicomponent liquids can be used to identify mixtures using the so-called fingerprint method. The result of the work of the "electronic tongue" can be a chemical image of the analyzed liquid or the so-called "fingerprint" obtained as a result of the integral evaluation of the solution. Comparing the "fingerprint" of the liquid under study with the "fingerprints" embedded in the system during system training, it is possible to identify the analyzed solution with a high degree of probability.

The most common application of multisensor systems of the "electronic tongue" type in the food industry is in the field of food analysis, both for monitoring and determining the concentrations of the main components in biotechnological processes of food production, and for determining the compliance of products with a given standard. The latter application enables the detection of counterfeit food products. The ability of the "electronic tongue" system to detect key substances that determine taste, with the issuance of information about the type and intensity of taste, allows them to be used when tasting unknown products. In addition to food, the artificial definition of taste is extremely in demand in the pharmaceutical industry.

A rapidly developing group of multisensor systems used for military, security, and other purposes are multisensor surveillance systems (see Figure 2.7).

Most advanced multisensor surveillance systems include a wide variety of sensors (Liu, 2010). Of course, the main sensor is the visible range color camera. In addition, depending on the tasks to be solved and the observation conditions, multisensor surveillance systems can include one or more thermal imagers, cameras for working in low light conditions, cameras in the near infrared region, short wave infrared or short-wave infrared (SWIR) cameras, a laser rangefinder, laser backlight and pointer, and some others.

Methods and algorithms for data processing are important in the operation of multisensor systems. In video surveillance systems, processing can consist of several stages, as shown in Figure 2.8.

At the stage of receiving data, preliminary processing of the incoming information takes place. The data fusion stage is designed to improve the quality of information. The pixel level segmentation step is designed to isolate objects of interest. At this stage, the methods of binarization of images of the selection of contours of objects are described in Bilan (2014). The fusion stage of pixel level segmentation combines images from different sources (for example, in the visible range and in the infrared range). At the stage of detecting deviations, information is filtered in order to select the most useful. Levels of identification and classification make it possible to recognize an abstract object and establish its correspondence to real objects. The stage of

FIGURE 2.7
Multisensor surveillance systems. (Picture taken from the site https://www.rospribor.com)

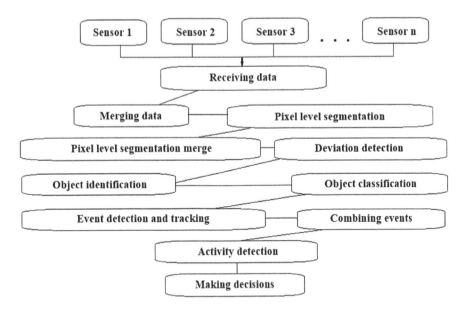

FIGURE 2.8
Stages of data processing.

detecting and tracking events allows you to identify the behavior of an object in space. Combining events allows you to conclude about the activity of the object. Activity detection phase allows you to judge the behavior of the object in real time. These stages, and even more so their sequence, are given as an example and do not claim to be the ultimate truth.

A striking example of multisensor systems is multisensor fire detectors. Such systems contain up to four sensors, because associated factors of any fire are smoke, temperature, carbon monoxide concentration, and open flames. Simultaneous monitoring of all four factors multiplies the reliability of such a detector. An example of a multisensor fire detector is a four-sensor device from System Sensor, shown in Figure 2.9.

In the past few decades, multisensor technology has developed rapidly. In this regard, it becomes necessary to classify multisensor systems. Depending on the classification criteria, all multisensor systems can be divided into five large groups (see Figure 2.10).

Classification by purpose defines the area of human activity in which multisensor systems are used. These areas can be industry (for example, robotics), medicine (diagnostic systems), military (detection and targeting systems for high-precision weapons), security (surveillance systems), and many others.

In terms of the volume of tasks being solved, multisensor systems can be local ("smart" home, "smart office"), regional ("smart" city, "smart area"), or global ("smart" country, "smart" planet, "smart" space).

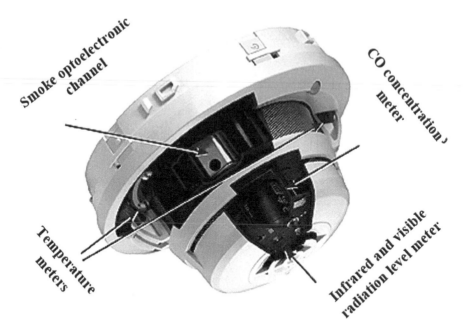

FIGURE 2.9
Four-sensor device of System Sensor company. (Picture taken from the site http://www.tzmagazine.ru)

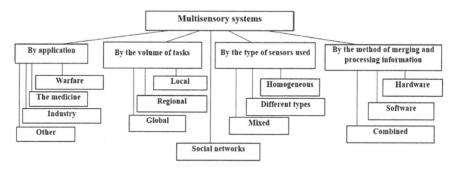

FIGURE 2.10
Classification of multisensor systems.

By the type of sensors used, multisensor systems can be classified into the same type, when sensors of the same type are used to control homogeneous parameters, different types, when the set of sensors is sensors for monitoring various environmental parameters, and mixed, when sensors are used together that not only control the same environmental parameters, but also sensors based on different principles of operation and designed to control different physical quantities of the surrounding world.

According to the method of merging and processing primary information, multisensor systems can be classified into hardware systems, when the data coming from sensors is fully or partially processed by hardware by the sensors themselves or by intermediate equipment (various types of converters) according to a rigid algorithm; software, when the algorithm for converting and merging data can be programmed and changes depending on the problem to be solved and combined, or representing a hybrid of hardware and software systems for processing and merging data (Buimistryuk, 2014; Esbensen, 2003).

A separate group of multisensory systems, quite new, associated with the informatization of modern society are social networks and messengers. Social networks are a special kind of multisensory systems. I will take the liberty of calling this phenomenon the multisensory systems of human souls. Such systems are associated with the processing of big data, which are distinguished by large volumes, high speed of filling, and a wide variety of types, which imposes increased requirements on hardware processing devices and information storage facilities, and also requires constant improvement of software data processing algorithms.

2.4 Conclusion

As a result of the analysis of sensor systems, it can be concluded that the increase in the capacity of computer technology, the development of artificial intelligence, informatization of society, automation of production, and the development of robotics give a powerful impetus to the accelerated evolution of precisely multisensor systems. In the work, the fact of a huge variety of multisensor systems was stated and an attempt was made to classify them. The most developing areas, at present, are multisensor systems that solve global problems in the field of management and informatization of society, as well as artificial intelligence.

The emergence of new open source platforms for storing and analyzing big data has given impetus to the development of multisensory systems for human souls. Filling with content of big data is carried out not only when using social networks, instant messengers, search engines, or internet commerce, but also with the advent of the Internet of Things, data is received on the models of user actions and the operation of products.

3

Smart Cities Based on Multisensor Systems

3.1 Introduction

Modern society is arranged in such a way that people live both separately and unite in the form of colonies. Most often, people form colonies for different reasons. Colonies can be large or small, as well as primary and secondary (Krichevsky & Dubovskaya, 2009). Primary groups are characterized by close interaction of all their members on the basis of psychological compatibility, intimacy, sympathy, and spirituality (family, friends, and other groups). Secondary groups are characterized by activities to achieve a common goal (employees of one organization, political party, etc.).

All human colonies are characterized by common features: territorial, linguistic, targeted, economic, spiritual, etc. Modern society is characterized by the rapid transition of individuals from one colony to another. However, there are colonies that are very stable, which are characterized by rigid and unchanging relationships. These relationships practically do not change; however, the colonies themselves can be very flexible and rebuildable when solving various social problems. It is customary to call colonies a certain group of people with different goals and characteristics, located on a certain territory and united primarily by common characteristics and pursuing the goals described earlier. The most important goals are the arrangement of any territory for protection and survival.

In recent years, society has been characterized by a high rate of urbanization of the population (Yamagata, 2020). Modern cities are increasing and the number of cities in countries becoming more (Picon, 2015; Song, Srinivasan, Sookoor, & Jeschke, 2017). People in such cities live in one colony and obey the laws laid down for the people of this city. All residents of such cities strive for a comfortable life. To improve the standard of living in the city, the city is trying to make it "smart".

According to McKinsey, 600 cities on the planet will become smart by 2020. What are smart cities and who is ready to live in them?

3.2 The Main Characteristics of the Smart City

A smart city is a living organism and is an integral part together with its inhabitants. All smart cities differ in detail. These differences are due to climatic conditions, the contingent of people, manufactured goods, social conditions, the tasks of residents, etc. First of all, a smart city characterizes the degree of use of digital technologies and the degree of intellectualization of digital technologies (Picon, 2015; Song, Srinivasan, Sookoor, & Jeschke, 2017; Cardullo, Di Feliciantonio, & Kitchin, 2019; Barlow & Levy-Bencheton, 2018; Hassanien, Elhoseny, Ahmed, & Singh, 2019). It is believed that by 2030 more than 60% of people will live in cities.

To make the city smart, you must first solve the following tasks.

1. Optimization of the transport system.
2. Energy saving and optimization.
3. Simplification of all systems of service to people.
4. High security.
5. Improving comfort and living standards.

Optimization in the transport system consists in the formation of transport routes, identification of "bottlenecks", identification of the minimum path, as well as the minimum route time, taking into account the congestion of highways. This section also includes the tasks of finding free parking spaces in real time on the territory indicated by the user.

Energy saving and optimization consists in the automatic distribution of energy in all urban areas of consumption, efficient use of various sources of accumulation, and storage of various types of energy.

Simplification of all systems of service to people solves the problem of digitalization of all types of payment services, as well as services in the identification of residents and guests of the city. There is implementation of electronic document management in order to simplify the procedures for servicing the population. Particular attention is paid to identifying individuals at significant distances.

High security lies primarily in high digital security, since smart city is mainly built on highly developed digital technologies with elements of artificial intelligence. High-tech data transmission and exchange systems must be used. Unauthorized changes in transmission protocols and information content lead to the destruction of all or part of the city's functioning system.

An increase in comfort and living standards is determined by optimizing the time and various costs spent on the functioning of a person and the entire city, as well as intellectual assistance and replacement of many physical actions that a person needs to perform (for example, replacing many sequences of actions for washing dishes, cleaning vegetables, and cooking).

A smart city takes over a number of human intellectual functions. All infrastructure and all functions of a smart city are aimed at helping people and at their effective protection. A smart city is an inextricable unity with its residents, even when residents are temporarily absent from it. In fact, a smart city assesses the psychological, physical, and intellectual state of residents and provides an automatic artificially created replacement for many cumbersome human functions.

What functions can a smart city perform?

1. Primary identification of a person.
2. Determination of the biological state of a person.
3. Determination of human needs in general and at the current time.
4. Performing the necessary control tasks coming from a person in automatic mode.
5. Replacement of a number of human intellectual functions and their implementation.
6. Protection of a person and his property.
7. Analysis of the state, infrastructure, transport, energy consumption, climate, as well as digital communications.
8. Prediction of various events both within the city limits and outside the city.

The primary identification of a person consists in the implementation of methods for identifying each person in an automatic mode. There are many methods and means for identifying a person. The most effective approach is to identify a person by analyzing their biometric characteristics. Identification by special identifiers that accompany a person within the urban environment is also used. If identification is carried out over significant distances, then different methods of digital and biometric identification are used. In this case, the most effective is biometric identification in real time, i.e. by dynamic biometric characteristics.

Determination of the biological state of a person is carried out using special primary converters (sensors) that determine body temperature, pressure, alcohol content, and other parameters. The received data is processed by special intelligent electronic means, which make a decision about the biological state of a person and then take this data into account to organize the functioning of the urban infrastructure. This data is also sent to the person, as well as to specialists (doctors, psychologists, etc.), to clarify the person's condition.

Determination of needs is to help implement a person's plans (for example, a reminder of an appointment, implementation of breakfast by serving food and serving the table, serving a car at a specified time and place, etc.). Also, the list of these functions includes the correction of the biological state of a person based on the information received from the decision-making centers about the biological state of a person.

Performing the necessary control tasks coming from a person in an automatic mode realizes the task of performing various payment functions and other everyday functions (for example, turn on a robot vacuum cleaner, TV, air conditioner, etc.). To perform such functions, a person must have access. For this, each resident uses passwords or applies his own biometric characteristics, the quantitative values of which are transformed by biometric sensors and passwords are formed.

Replacement of intellectual functions and their implementation is carried out in all areas of activity, covering the city-wide scale, individual houses, and apartments. At the city-wide level, optimization of traffic is carried out, as well as the distribution of energy resources, assistance in finding the optimal paths of movement for traffic objects in the city, street lighting, houses lighting, etc. At the level of individual houses, automatic comfort is provided for all residents. When approaching the house where a person lives, comfort begins to be felt in primary biometric identification, automatic door opening and other sequential comfort operations.

Protection of a person and his property is based on the cybernetic protection of all his personal data, as well as the protection of information with which every resident of the city operates. In addition, the protection functions include the implementation of secure access to the home, to various institutions, and to various automatic means of functioning and control.

All digital communications analyze the state, infrastructure, transport, energy consumption, climate, as well as digital communications. This helps prevent emergencies and effectively manage the entire urban economy in an automatic mode.

The intelligent technologies used make it possible to predict the state of the entire infrastructure and, on the basis of this information, effectively plan the entire sequence of events and work performed by the main automatic means implemented in the urban area.

All communications can perform unplanned actions that are implemented as a result of intelligent decisions due to the behavior of people inside the city, as well as due to external factors affecting people and infrastructure. In this regard, all information technologies use the possibilities of self-organization, aimed at solving new problems that were not originally incorporated in the digital architecture of the city. Self-organization entails a restructuring of the connections of all or part of the city's communication, as well as redistributing local operations for each computing module or organizing new computing modules.

All information about events in the city comes from primary converters (sensors), which convert analog signals of various nature into signals that are not perceived by the observer. These signals from the sensor outputs are converted into a digital signal, which is further processed by a computer system. It is considered a good approach in the implementation of a smart city, which uses sensors that generate a signal at the output of the same nature. Also, sensors do not have to be expensive and complicated. They should be easily understandable to users and provide a quality measurement and reading of all necessary information within their scope.

3.3 Biometric Sensors in a Smart City

Among all the sensors used in a smart city, a significant place is occupied by sensors aimed at the perception of biometric information about a person (Fairhurst, 2018; Boulgouris, Plataniotis, & Micheli-Tzanakou, 2009; Bilan, Elhoseny, & Hemanth, 2021c; Peng, Favorskaya, & Chao, 2020). To record biometric information, sensors are usually used that convert optical and acoustic signals. The most commonly used sensors are video detectors (video cameras) and microphones. Video cameras capture images of various parts of the body, and acoustic cameras capture a person's voice in the form of digital codes.

All recorded signals are transmitted to the block for processing and making decisions about identification. These blocks carry out the hardware and software implementation of identification methods based on the selected biometric characteristics. The implementation of the methods can also be carried out directly in the sensors themselves, and the identification code will already be transmitted. This improves performance and provides real-time scale.

Along with identification, sensors are used that record various physical characteristics of a person, which determine his physical state. Such sensors have different design principles and are not optical or acoustic. They aim to respond from a wider range of analog signals.

All other functions of a smart city are implemented on the basis of information received from sensors or from the person himself. These functions are carried out by means employed by people. These means are developed in advance, implemented and the formation of network communication and intellectualization of the entire infrastructure of the city is carried out.

From the totality of sensors, sensor networks are formed, which are critical components of smart cities, since the data they collect is fundamental to these services (Ferrari, 2010; Krit, Bălaş, & Elhoseny, 2020; Peng, Favorskaya, & Chao, 2020). Sensor networks provide an automatic mode of city functioning.

It is important to use wireless data transmission within the city. Currently, such a regime is provided using many modern technologies.

Wireless sensor networks (WSN) can be built with simple and cheap sensors with low power consumption and cost and do not require expensive maintenance. WSNs allow continuous monitoring of the urban environment and the surrounding area.

A smart city integrates information technologies, communication systems, and intelligent technologies to make decisions of various kinds. Computing and WSN integration enables real-time data mining and decision-making, giving the ability to see different situations in a smart city.

4

Biometric Characteristics

4.1 Static Biometric Characteristics

Biometrics is a science based on describing and measuring the characteristics of the body of living beings.

As applied to systems of automatic identification, biometric means those systems and methods that are based on the use of any unique characteristics of the human body for identification or authentication (Bilan, Elhoseny, & Hemanth, 2021c; Boulgouris, Plataniotis, & Micheli-Tzanakou, 2009).

Biometric identification is often called pure or real authentication since it is not virtual, but actually related to a person's biometric sign (identifier) that is used. This identifier is displayed by the quantitative characteristics of each trait of a person.

To implement networks for collecting and processing information, many biometric parameters are used. They are divided into static and dynamic. Biometric characteristics are common to all people. Today, many biometric characteristics are open and described. For example, for the geometric characteristics of the palm image, more than 70 biometric parameters have already been described and new ones are being discovered in the form of individual geometric parameters and their ratios. There is an intensive search for new biometric characteristics that require accurate, uncomplicated, and cheap methods and tools for their processing.

Static biometric characteristics include those characteristics that are constant over time. They are fixed at a certain point in time and are unchanged. Static characteristics include: fingerprint, images of the face, palm and ear, iris, pattern of veins on different parts of the body, image of the retina, facial thermogram, DNA and others.

Dynamic parameters include: handwriting and signature dynamics, heart rate, voice and speech rhythm, gesture recognition, speed and features of keyboard work, gait, etc.

The most common biometric parameter is the fingerprint. The fingerprint is presented as a skin pattern image of the finger. For each person, he is unique. An imprint with the help of a scanner forms an image of a drawing, which is processed, described, and converted into a digital code, which comes to the unit for comparison with the standards generated by the user earlier. This parameter can be read using a special scanner, which requires a tight finger to the surface of the scanning device. Difficulties arise when the finger is located far from the sensor. Therefore, this parameter is used only by consent or at the request of the user who needs to be identified. The most common fingerprint sensors are optical scanners. However, they are not resistant to false images and respond poorly to a real "live" finger. For example, scanners used in mobile devices cannot clearly scan a fingerprint of a wet hand.

Iris identification and authentication is also device-dependent and requires the user's consent and desire for such identification. This technology is aimed at identification by the unique characteristics of the human iris. An image is processed with circles and patterns that make up a unique pattern for each person. More than 70 information points are highlighted in the figures, which make it possible to identify the user. The methods and means used are quite complex and require fixing the human head in a certain place in relation to the input aperture of the scanning device. It is practically impossible to read the pattern of the iris when a person moves freely in space and is at a considerable distance from the sensor.

Retinal identification and authentication uses an image of the fundus blood vessels. Sensors use weak infrared radiation to capture human patterns. This technology relies on means of highly accurate pattern reading. Expensive high-precision equipment is used, which requires the close position of the eye to the scanning device. This technology is also device-dependent and psychologically affects the identified person.

Hand geometry authentication and identification is considered the simplest method. However, it also requires a fixed position of the hand on the surface of the scanning organ. Several geometric characteristics are defined, such as finger lengths, curves, finger thickness and width, and other geometric characteristics. The image of the hand is quite simple and does not require complex processing methods. At the same time, the location of the fingers affects the identification result (clenched or unclenched fingers, bent fingers). Therefore, additional fixing elements are located on the surface of the scanner. The identification process is well automated and widely used. In modern methods of biometric personality identification based on the geometry of the hand, a fixed part of the hand is used, which gives a more reliable and stable identification result (Bilan, Bilan, & Bilan, 2021a; Yuzhakov, Bilan, Bilan, & Bilan, 2021).

An ear biometrics system can be thought of as a typical image recognition system that describes an input image using a set of basic geometric characteristics and compares it with a base of other sets to establish an identity. Ear recognition can be implemented for a flat method or a three-dimensional spatial set of points depicting the surface of the external auricle. This method can be implemented at a considerable distance from a person and does not require his consent. In addition, biometric identification can be carried out by the inner shape of the auricle. In this case, special devices are used, which are inserted into the auricle and scan its surface.

Face geometry identification is now gaining more and more popularity. However, the facial image has a number of differences. Therefore, a large number of geometric components are used, which together give reliable differences between one person and another. This parameter, depending on the method, does not always require a clear fixation of the person's face in relation to the reading device. The methods of such identification are rather complicated and have not yet acquired a universal approach. All of them are based on the specific developments of researchers of this problem. Smiling, sadness, squinting eyes, or moving eyebrows can distort the identification process and lead to a false result. This also applies to turning the face in relation to the video camera, which sharply complicates identification methods. A person easily recognizes any caricature of a person he knows, which until now cannot be done by any computer.

Identification and authentication of a person by the image of a handwritten signature is carried out by analyzing the image of a person's personal signature reproduced on paper or using special tablets. This method cannot always give an accurate identification result; therefore, it is used as an additional characteristic in multimodal methods of processing biometric information. This parameter is used in banking and is often used as an additional access protection parameter.

Vein pattern authentication is a relatively new biometric method and is highly reliabile. The use of a vein pattern as an identifier for a biometric system demonstrates a high level of reliability with a small number of failures: false acceptance rate (FAR) – 0.0008% with false rejection rate (FRR) – 0.01%. Even with a template, it's extremely difficult to create a dummy. The method is also device-dependent and requires a special arrangement of selected body parts with vein patterns (finger, palm, etc.). This method is applicable only with the user's consent for identification. Vein pattern readers are sensitive to halogen light or direct sunlight, which creates special requirements for the lighting conditions at the scanner installation site. There are diseases that disrupt the pattern of veins, which leads to false identification or non-identification.

Static biometric characteristics are used everywhere and are being introduced at a rapid pace in almost all areas of human activity.

4.2 Dynamic Biometric Characteristics

Biometric authentication and identification methods based on dynamic biometric features are based on the behavioral (dynamic) characteristics of a person. These methods analyze the features characteristic of subconscious movements in the process of reproducing an action. They take into account the psycho-physiological characteristics of each person.

A simpler and more commonly used method based on dynamic biometric characteristics is voice identification. A microphone and equipment are used to convert sound signals into electrical signals and process them. Combinations of frequency and static voice characteristics are taken into account. At the same time, the method has low accuracy, so a person can have a cold or have a different mood, which can affect his voice combinations. Also, for identification, a person must speak and not shout, and not speak in a whisper. In addition, during the identification process, various sound signals from other sources may occur, which distort the useful signal. Therefore, there is a need to introduce isolation into the receiving device or embed frequency bandpass filters.

Identification by the dynamics of handwriting is based on taking into account the movements of a person's hand when he reproduces handwritten text on paper or on a special tablet, the surface of which is susceptible to pressure. Plates that are sensitive to pressure allow such dynamics to be described in three dimensions. They use different principles for constructing a react surface. Values generated during signature reproduction are processed and a database of standards is formed, which participates in the identification process.

Identification by keyboard handwriting is based on the analysis of the time characteristics of keystrokes and key retention. Using both hands, each person creates an individual approach to typing. The individuality of the user relies on the speed of typing, different habits about keystrokes, etc. Such features have made it possible to create a number of biometric identification methods based on the dynamics of handwriting working on the keyboard. For biometric identification by this parameter, fixed and arbitrary key phrases are used. The method allows to determine the psychological state and state of human health. A simple program can be implemented to test the authentication capabilities with keyboard handwriting. The infusion rate will float within a certain range over a period of time. The method does not require any additional actions from the user. The user still enters his password when logging in, based on this password, additional authentication can be implemented. The advantage is the relatively simple implementation of the method. There is also the possibility of hidden authentication – the user may not suspect that he is undergoing an additional check for keyboard handwriting. Moreover, the method has a strong dependency on a particular keyboard. If the keyboard is replaced, the user needs to configure the program again. The method is dependent on the psychological state and health of the user. For example, if a user is sick, they may have different keyboard handwriting options. The method can be applied for remote identification in network applications.

To date, such a biometric characteristic as gait has been little studied and there are no high-precision methods and means of biometric identification for this parameter. In this direction, modern developments are actually marking time in one place in development and improvement. Therefore, for many reasons, the task of biometric identification of a person by the dynamics of his gait is one of the first tasks in many automated systems. A person's gait is analyzed for a certain period of time, and the result is the study of the dynamics of changes in quantitative characteristics over time. Such dynamics is individual for each person. Various methods are used to analyze gait.

The widely used methods are based on the use of special sensors that are attached to various points of the human body. Information is taken from them, which is further processed. This approach is highly hardware dependent and often uses a special coating on which a person can be identified. Of course, this is not a solution to the problem, since on different soils (asphalt, grass, sand, etc.), a person walks in different ways. Therefore, it is better to use a remote method based on the use of video sensors and the formation of a video sequence, which is then processed and a decision on identification is made. However, for such an approach, the methods are quite complex and require special, accurate, and intelligent image processing methods.

The uniqueness of the electrical activity of each person's heart prompts the use of an electrocardiogram as a biometric parameter in various security and authentication systems due to the ease and cheapness of signal withdrawal, as well as the complexity of its forgery and involuntary extraction. At the moment, various approaches are used to study the possibility of identifying a person by ECG. The identification mode includes the following stages: data collection, processing, extraction of characteristic features, and classification. At each of these stages, research teams use different mathematical algorithms: principal component analysis, wavelets, neural networks, etc.

As mentioned earlier, a large variety of biometric characteristics are used and they can be considered for a long time. The most popular characteristics were considered.

4.3 Multimodality of Biometric Identification

However, among all biometric characteristics, it is difficult to talk about high reliability of identification by one parameter. Therefore, a more reliable method of biometric identification is the use of multimodal models. For this, several biometric characteristics are used, which are processed simultaneously. For example, using a mobile phone, you can simultaneously analyze a fingerprint image and an image of a person's face or the shape of his ear. This method provides high reliability of identification. It is important to select the necessary biometric characteristics, which are simply perceived and processed by sensors.

A number of quantitative characteristics are used to evaluate the biometric identification system. The main characteristics are:

- FAR – false recognition rate;
- FRR – false non-recognition rate;
- Receiver operating characteristic(ROC) curve – graphical relationship between errors of the first and second kind;
- The equal error rate (ERR) is the rate at which both errors (receive error and rejection error) are equivalent. The lower the EER, the higher the accuracy of the biometric system.
- False match rate (FMR) – the probability of a false match of parameters. In this case, one sample is compared with many templates stored in the database, i.e. identification occurs.
- False non-match rate (FNMR) – the probability of a false mismatch of parameters, in this case, one sample is compared with many templates stored in the database, i.e. identification occurs.
- Curve (DET) – graph of probabilities of comparison errors (FNMR versus FMR), probabilities of decision errors (FRR versus FAR) and identification probabilities on an open set.

It is desirable to consider all these parameters together. The presence of ROC and DET charts indicates high confidence in biometric system research. The FAR and FRR parameters should be small. The smaller they are, the better the biometric system is.

Biometric characteristics can be divided into characteristics, which are defined as hidden and open. Hidden biometric characteristics are invisible to other people. By these characteristics, people do not identify each other; however, they often use them for accurate identification. These characteristics include: an image of a fingerprint, a drawing of veins on a finger, the dynamics of handwriting, the dynamics of handwriting on the keyboard, DNA, etc. These characteristics are determined using special methods and tools that the person himself develops.

Open biometric characteristics do not require special means to identify a person by a person. However, a person often makes mistakes when identifying and therefore additional hardware and software are used to help a person in biometric identification based on open biometric characteristics. These biometric characteristics include the image of the face, the iris of the eye, the shape of the ear, the shape of the palm, the image of an individual handwritten signature, gait, smell, voice, etc.

Also, biometric characteristics can be divided into characteristics that can be analyzed at a distance and in close proximity to the input of the biometric sensor. All hidden biometric characteristics and some of the open ones refer to biometric characteristics that require the immediate proximity of the biometric sensor.

The number of biometric characteristics that can be analyzed at a considerable distance is quite small. These characteristics include: geometric shape of the body, geometric shape of the face, geometric shape of the ear, smell, gait, etc. Basically, such characteristics require the use of video sensors (video cameras), since they process optical information and can do this at a distance.

Currently, a search is underway for new biometric characteristics that would simplify the identification process and reduce the percentage of false identification.

It should also be noted that there is a whole set of additional biometric characteristics that do not appear immediately after birth and can be determined using special methods and means. Such characteristics are called acquired biometric characteristics. They manifest themselves in the process of doing some kind of work or living in certain conditions. For example, the dynamics of the handwriting of working on the keyboard does not appear immediately, but only after working on it for a long time. A person acquires individual skills, which are determined by his intellectual qualities and physical condition, as well as his attitude to the instrument used.

Another example would be the dynamics of a shovel when digging a trench. Also, many different examples of working with any tools are possible. The individual physical and intellectual properties of each person are manifested here. Each person's handwriting is also an acquired biometric feature, as it is obtained as a result of working with a writing element. If there were no writing element and there were no teacher, then the person would not have such a biometric feature.

In addition, biometric characteristics include characteristics that show only the intellectual properties of each person. These characteristics include such as the choice of a picture from a set of pictures, an arbitrary text sequence formed by a person, and other characteristics.

To increase the identification accuracy, the best solution to the problem is to use multimodal methods for processing biometric data.

5

Biometric Data Processing in Smart Cities Based on Multisensor Systems

5.1 Uniform City Biometric Community

Today there are no smart cities that perform the full range of functions that satisfy all the requirements of the residents of this city. Also, such smart cities will not appear in the near future. As mentioned in previous chapters, biometric data processing (BDP) occupies an important place in the organization of a smart city.

What are the functions of BDP?

The answer is to define the tasks of BDP. There are two main objectives of BDP.

1. Personal identification and authentication.
2. Controlling human access to physical and virtual objects.

These problems are ubiquitous, even outside cities. However, solving these problems does not make it possible to implement a set of intellectual functions inherent in a smart city, where a unified citie biometric community (UCBC) is organized.

UCBC – a system of interconnected biometric data of a city. This system is very flexible. It is constantly changing since some biometric data that was already present in it can be transferred to the archive at any time (in connection with moving to another city or death), and new biometric data can also be added in connection with the arrival of new residents from other cities or the birth of new residents.

Data links can be implemented in various ways. The main task of such a structure is the virtual creation of communities based on biometric data. In such a system, communities can be organized according to the properties of belonging of a city resident.

Each community (set) of people in such a system organizes a biometric set. For example, the same biometric data can be composed of many residents of a house and many employees of the same company. This approach creates a serious and complex interweaving of connections, the intersection points of which can give out a new quality and form new sets with intersecting properties.

The structure of such a system is shown in Figure 5.1.

Biometric characteristics (BCs) from city residents are distributed over biometric sets through the distribution system of BCs over sets. The system contains various sets, which are determined by the selected BCs when designing the system.

The sets that are defined by one BC are usually larger than the other sets and define groups that do not require a high degree of protection. Such groups may include, for example, residents of the same house or neighborhood. The structure of the UCBC also includes sets determined by several BCs. These sets are more reliable and protected from unauthorized access. Such sets can unite residents who work in the same enterprise or office. Also, these sets can be formed by the intersections of residents of two other sets or more.

The formation of sets that characterize the intersections of various sets that are not specified initially is of great importance. An example of a set formation diagram is shown in Figure 5.2.

The diagram shows an example of the formation of sets based on three BCs for four people. Sets are formed, defined by combinations of BCs. All sets are different; however, there are sets that are formed taking into account a

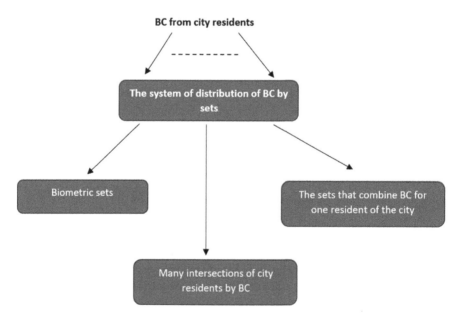

FIGURE 5.1
UCBC structure.

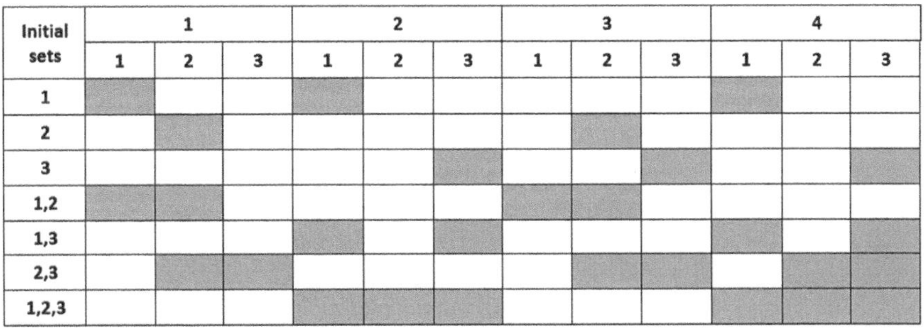

Initial sets	1			2			3			4		
	1	2	3	1	2	3	1	2	3	1	2	3
1	▓			▓						▓		
2		▓					▓					
3						▓			▓			▓
1,2	▓					▓					▓	
1,3				▓					▓	▓		▓
2,3		▓						▓			▓	
1,2,3				▓						▓		

FIGURE 5.2
Diagram of formation of sets.

large number of BCs. For the example under consideration, the set, which is determined by all three BCs, unites the second and third person. In this case, the intersections of the sets of these people are determined by other BCs. For example, sets 1; 3; 1.3 also cover the second and third person. However, these sets may include other residents of the city. As a rule, sets consisting of a smaller number of inhabitants are determined by a large number of BCs. For example, residents of one house define a set of one BC, residents of one floor define a set with two BCs, and residents of one apartment define a set with three BCs. Such sets can be clearly and visually represented graphically. Such sets are homogeneous since they have the same property, which indicates the location of a city dweller.

In the case of the formation of sets according to the properties of professional belonging, then such sets will cover the entire territory of the city and will partially intersect with sets determined by the properties of the location. Such sets can be displayed graphically. In general, all sets will be indicated by different colors and dashed lines (Figure 5.3).

Sets are represented by vertices, and an example of intersection of sets with common properties is presented. A program has been developed that allows you to determine the intersection of sets by given biometric properties. For example, 11 subjects were created, among which BCs were distributed. City residents are described by two to five BCs. An example of city residents who have the fifth property and belong to a certain group is shown in Figure 5.4.

It is assumed that all residents recorded in the program accidentally move around the city. As you can see, different BCs are assigned to different residents, which assigns them to different sets, and at the same time, some of them have the same intersection of characteristics that unite them by certain properties.

If several BCs from the considered examples are taken into account, then the sets sharply decrease (Figure 5.5). However, it is possible to determine the belonging of an inhabitant to the selected sets.

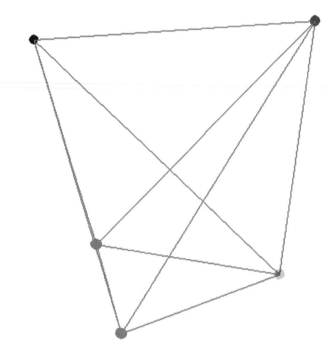

FIGURE 5.3
Example of representation of sets and their intersections.

The system is characterized by the fact that sets can change, and new sets can also form, aimed at assisting in solving intellectual problems within a smart city. It is important to determine the required intersection of the sets. For each city, the necessary intersection of sets should be determined, which would create the most comfortable conditions for city residents. It also identifies critical set intersections that can lead to disease and other problems in urban infrastructure.

For a broader and more accurate definition of such sets, a program has been developed that uses anthropometric and biometric characteristics. A database is formed that includes various identification data. These include codes, passwords, physical quantitative characteristics (height, weight, etc.), and codes of various BCs. On the contrary, this program forms sets according to the quantitative values of anthropometric and biometric characteristics. The program also includes the use of the coordinates of the location of city residents throughout the territory, or in separately adopted sectors into which the city's territory is divided. An example of the initial location of city residents is shown in Figure 5.6.

Based on the obtained distribution, various sets are determined by the biometric indicators of residents. For this, the program includes filtering

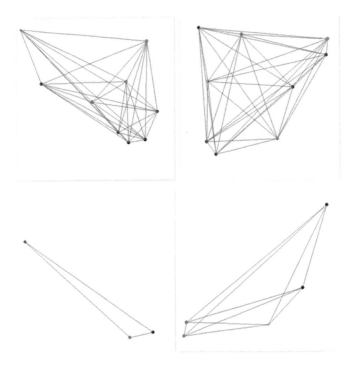

FIGURE 5.4
An example of many residents of a city belonging to the same group.

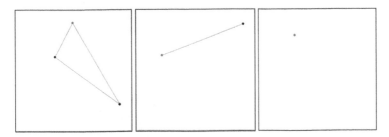

FIGURE 5.5
An example of sets containing several identical biometric characteristics.

by many quantitative indicators. For example, according to the quantitative characteristics of human growth, the distribution has the form shown in Figure 5.7.

In Figure 5.7, the set of all residents is highlighted, whose height is more than 180 cm.

The next filtering parameter is the weight, the value of which is more than 60 kg (Figure 5.8).

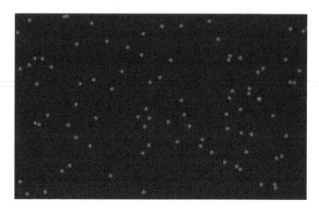

FIGURE 5.6
An example of the initial location of city residents.

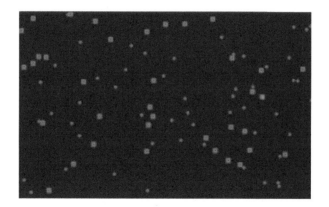

FIGURE 5.7
An example of a set obtained by filtering by the height of a city resident.

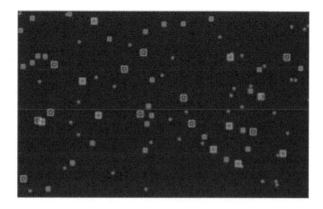

FIGURE 5.8
An example of a set obtained by filtering by the height and weight of city residents.

FIGURE 5.9
Example of sequential addition of biometric characteristics.

In Figure 5.8, new inhabitants are added by selection with large rectangles.

The program allows you to select sets by BCs. An example of the sequential addition of BCs is shown in Figure 5.9.

Figure 5.9 shows the sequence for adding BCs from left to right. Moreover, each added BC is also amenable to filtering.

Changes to the original sets constantly lead to changes throughout the UCBC web. Knowing the desired and critical sets, you can manage the best location of residential buildings and the location of enterprises, hospitals, and other institutions.

5.2 Biometric Features in the Smart City

As mentioned earlier, BCs in a smart city are divided into two types according to the method of fixation.

- BCs requiring close contact with biometric sensors;
- BCs that recorded at a considerable distance.

The first type of BCs is used mainly in access systems, and the second type of BCs is used in surveillance and identification systems. For a smart city, the second type of BCs is suitable, which do not require time and human attention to solve an intellectual problem by the system. These characteristics include: face image, geometric shape of the ear, geometric shape of the palm, gait, and other characteristics. Also, in a smart city, you can talk about BCs, which are determined using sensors that react to chemical influences (for example, smell). These characteristics can be determined from a distance.

5.3 Human Face Image Analysis

Currently, a lot of attention is paid to biometric identification of a person by the image of his or her face (Saračevič, Elhoseny, Selimi, & Lončeravič, 2021; Abidi & Abidi, 2007; Boulgouris, Plataniotis, & Micheli-Tzanakou, 2009).

Many methods and tools have been developed that have high results (more than 95% correct identification). Most of the methods are implemented over a considerable distance. In this case, a person may not know that he or she is being identified. This approach is implemented using optoelectronic means of capturing the image of a person's face. Existing methods and tools allow you to effectively process images of faces obtained at different angles of the video camera (Figure 5.10). The images shown in Figure 5.10 are sourced from the free ORL database (AT&T Database of Faces 'ORL Face Database' AT&T Laboratories, Cambridge).

Most of the methods use various control points on the face image and determine their locations, as well as the distances between them (Saračevič, Elhoseny, Selimi, & Lončeravič, 2021; Abidi & Abidi, 2007; Boulgouris, Plataniotis, & Micheli-Tzanakou, 2009; Bourlai, 2019). However, in the three-dimensional format, a large number of standards are used for identification, which must be compared with a set of characteristic features at the input of the identification system. Various distortions in the form of smiles, blinks, etc. are also taken into account. Large amounts of data are used to process images of faces.

Now such systems may well be used within a smart city with limited video surveillance sectors. In this case, there should be several cameras for each

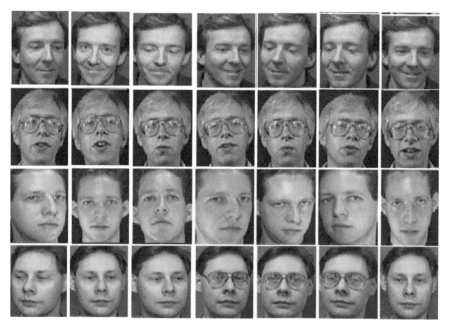

FIGURE 5.10
Images of faces taken from different angles of the camcorder.

sector, so that the image of the face of the identified person in any location in the space of the sector is recorded in a sufficient volume (directly or in profile). The system is not able to identify a person from the image of the back of the head.

5.4 Analysis of the Geometry of the Auricle

Another BC is the image of the human auricle. The auricle is an integral part of the image of a person's face and is involved in the processes of personality identification based on the geometry of faces. However, the auricle is an independent BC that can identify a person.

There are many methods that successfully identify a person by the geometry of the human auricle (Motornyuk, Bilan, & Bilan, 2021; Boulgouris, Plataniotis, & Micheli-Tzanakou, 2009; Nejati, Zhang, Sim, Martinez-Marroquin, & Dong, 2012; Afolabi & Ademiluyi, 2015). For this, special sensors and optoelectronic fixation devices (video sensors) are used. Currently, sensors are used to analyze the internal structure of the human ear. They use direct contact, which is not always pleasant for a person. However, they give a high percentage of correct identification.

In a smart city, most often the contact method of biometric identification by auricle is not acceptable since the implementation of a contact can take a lot of time. In a smart city, the non-contact method is more acceptable, which is implemented using photo fixation of the auricle. The resulting images are entered into a digital computing system, where they are processed and compared with the standard. For the processing of images of the auricle, special software is being developed that implements specially developed methods.

The disadvantage of both BCs is the presence of long hair on the head, which can cover large areas of the face and ear. There are hairstyles that cover the entire ear area. Examples of images of the auricle are shown in Figure 5.11. These images are freely available on Internet sites, as well as recorded by the authors themselves for different camera positions. These images were freely available from Esther Gonzalez.

The photo (Figure 5.11) shows several types of photos. They are formed at different camera angles and displayed against the background of the image of the rest of the head (hair, cheeks, forehead, etc.).

Photos of the auricle, which are selected with the help of special tools, do not require complex processing methods. However, they are possible only with the consent of the person being identified, which is not always possible in a smart city environment.

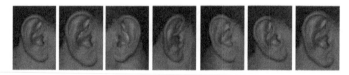

FIGURE 5.11
Examples of photos of auricles fixed at different angles.

Photos of the auricles against the background of the rest of the head contain additional elements in the image that must be ignored. To isolate the auricle, it is necessary to use specially developed algorithms implemented in software. These methods and algorithms are quite complex and require preliminary intellectualization when processing the resulting image. Intellectualization of the method consists in searching for an image of hair and other parts of the face that do not belong to the image of the auricle. Certain body parts are subsequently ignored by the method and software.

Thus, the main problem for contactless biometric identification based on the image of the face and auricle is the presence of long hair and headgear. This situation does not make these characteristics the main ones on the streets of a smart city.

Based on the preceding facts, it is necessary to look for BCs, which for the most part are independent of biometric sensors and can be recorded at a considerable distance. These characteristics include a person's gait.

5.5 Human Gait Analysis

The gait of the same person may be different when walking on different surfaces (asphalt, sand, grass, and other surfaces). Also, a person walks differently if he carries heavy objects (on his back, in his hands, etc.).

The most common methods for analyzing human gait are as follows:

- identification method based on a special contact surface on which a person walks;
- identification method that uses special contact sensors;
- identification method based on optoelectronic technologies.

The first method requires the use of a special sensitive surface, without which identification is impossible. A person must walk a certain number of steps on this surface. Therefore, such a surface should have a large area.

This method can be used in special rooms or areas and can claim universality. The following can be used as characteristic features:

the force of pressing the right and left feet on the touch surface;

the distance between the footprints of the right and left feet;

the time of holding down the right leg on the surface;

the time the left leg is held down on the surface;

the time of simultaneous touching of the surface of the right and left legs;

time between touches of the surface with the right foot;

time between touches of the surface with the left foot;

the distance between touches of the surface with the right foot;

distance between touches of the surface with the left foot.

Other characteristic features that are used for identification can be selected. The use of a special surface limits the method. Therefore, in a smart city, the method can be used in individual areas where it is very necessary.

The second method, using special contact probes, can be quite accurate. However, the method requires a lot of calculations and also uses a large number of sensors manufactured using special technologies. Sensors are attached to specific areas throughout the body.

With the help of sensors fixed on the body, temperature, pressure, and pulse rate at certain points of the human body are analyzed. Commonly used sensors are accelerometers, which measure the acceleration of movement. Also, gyroscopes, force sensors, bend sensors and ultrasonic sensors are used to analyze gait and form a set of characteristic features (Kazantseva, 2013). Different jobs use different numbers and combinations of sensors. Such designs and combinations are often used to create smart shoes and smart clothes. However, each sensor cannot be attached to any part of the body. Many sensors can only be attached to specific areas of the body. For example, accelerometers are most often placed on the waist, arm or wrist (Kazantseva, 2013), as well as in pockets and ankles.

Based on the use of such sensors, acceleration and horizontal movement of the leg in the plane are used as characteristic features. At the same time, different locations of the sensors give different results, which requires different approaches to computing.

Based on the use of such sensors, acceleration and horizontal movement of the leg in the plane are used as characteristic features. At the same time, different locations of the sensors give different results, which requires different approaches to calculations.

Many modern sensors are electronic and require power supplies. Also, electrical cables or radio air can be used for data exchange. These factors also limit the use of methods in a smart city environment.

In our opinion, the method using optoelectronic technologies is the most effective in a smart city environment. The method is implemented on the basis of a video camera, which can be located in different places in relation to the movement of a pedestrian (from the side, in front, from above, at different angles).

The video camera generates a video sequence of frames, which is further processed using specially developed methods, implemented in software and hardware. Different video sequences are generated for each location of the video camera. Accordingly, the following change: the gait period, the geometric dimensions of the body, and other characteristics. Different video sequences are analyzed using different methods. For example, a video sequence formed in front or behind a walking person should take into account the increase or decrease of the person's body over time, i.e. on every frame.

This method does not require specially developed technical means and can be implemented at a great distance from the pedestrian. In cities, most houses have solid walls that easily form the backdrop for a pedestrian. Video cameras can use different physical principles of video sequence formation.

In a smart city setting, human gait analysis can be successfully used for biometric identification. The method is implemented over long distances. A person may not know about it. However, at the moment, no gait analysis methods have been developed that would provide high accuracy of biometric identification. In addition, this method takes a certain amount of time to form a video sequence.

Based on the analysis, we can conclude that the best way to implement biometric identification in a smart city environment is a multimodal approach, which involves the use of several BCs. In this case, each used BC will complement each other, which significantly increases the accuracy and reliability of biometric identification.

6

General Information About Parallel Shift Technology

6.1 Theoretical Foundations of Parallel Shift Technology

Automatic image processing can be used in video surveillance systems, various security sensors, and machine vision systems. Video surveillance and machine vision systems can be part of both stationary and mobile objects. One of the options for constructing these systems is the use of parallel shift technology (PST) (Belan & Yuzhakov, 2013a, 2013b, 2018; Bilan, Yuzhakov, & Bilan, 2014b).

PST is the process of organizing the interaction of an object, the elements of which are described by n-dimensional vectors of characteristic features (CFs) with another object, the elements of which differ from the elements by the values of one or more CFs. This process is similar to the parallel shift of a set of elements of the initial set in some space of characteristics.

As such an interaction, we can use any methods and techniques that allow it to be detected. An example can be the use of any algebraic or logical function whose arguments are elements of the vectors of the CFs of the interaction objects, or directly elements of these objects. We can also use processing tools of sets for this purpose.

PST allows, firstly, to bring in a dynamic in the process of formation of the CF vector, if the characteristics are static, and, secondly, to reduce the total number of the characteristic features used for analysis.

The use of PST in image processing systems involves the manipulation of video stream frames. In the following examples of binary image processing, the interaction of objects is defined as the intersection of two images. The second image is a copy of the original image, the elements of which are parallel shifted in a plane relative to the elements of the original image.

Image processing by means of PST involves a certain generalization of information. Therefore, these methods are more productive to use to classify rather than identify objects of research. This method of information

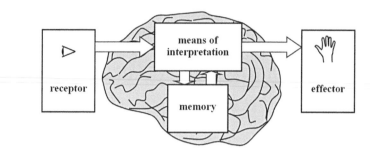

FIGURE 6.1
Scheme of functioning of the intelligent system.

processing is best used as part of the vision system of autonomous robots and automatic transport systems.

The intelligent system (IS) should include the following components: receptors, means of interpreting the received or stored information, means of storing information (memory), and effectors (means of interaction with the outside world) (Figure 6.1).

The principles of interaction with objects of external space for biological and automatic ISs are the same.

6.2 Data Generation and Storage

In previous works on image processing using PST (Bilan & Yuzhakov, 2018; Bilan, Yuzhakov, & Bilan, 2014b) noted that information that reflects a certain figure is stored in memory as a certain reference surface. This surface is formed by a set of the functions of the area of intersection for all possible shift directions (from 0 to 2π).

The function of the area of intersection (*FAI*) is a set of values of the area of intersection of a figure and its copy, which is shifted in parallel in a certain direction φ (Figure 6.2). The initial area of the image is denoted by S_0. The shift is made from the moment of complete coincidence of the figure and its copy until the moment when their intersection is equal to zero. This shift distance is called the maximum shift. In this case, the shift direction φ coincides with the direction of the OX axis of the Cartesian coordinate system. The maximum shift in this direction is denoted by X_{max}.

$$FAI(0) = S_0$$

$$FAI(X_{max}) = 0$$

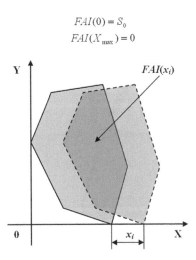

$$FAI(0) = S_0$$
$$FAI(X_{max}) = 0$$

FIGURE 6.2
The formation of the *FAI* when shifting a copy of the figure at a distance of x_i.

The values of the FAIs when shifted in opposite directions are similar. The reference surface has the property of central symmetry, so it is enough to store only half of its data in memory (a set of FAIs in the range of directions from some φ to $\varphi + \frac{\pi}{2}$).

Technically, the formation of *FAI* can be organized as follows. The initial image is located on a certain matrix of elements (the basic matrix – BM). For each given shift direction φ there is a certain shift matrix (ShM$_\varphi$), which is a set of cyclic shift registers. Shift matrices are physically inverted relative to the basic matrix by a certain angle (Figure 6.3). The number (n) of ShM$_\varphi$ is determined by the device developer. If their rotation relative to each other differs by an angle $\Delta\varphi$, then to ensure the formation of a complete reference surface, the following equation must be true: $(n + 1) \cdot \Delta\varphi = \pi$. The values of FAIs for directions 1 and $n + 1$ are the same due to the properties of central symmetry. The sizes of the BM matrices and each ShM$_\varphi$ matrix do not have to be the same. The main thing is that the images that are located on them were the same. However, for unification, we will assume that the sizes of BM are equal to the sizes of ShM$_\varphi$. We call the elements of the basic matrix $a_{i,j}$, and the elements of the shift matrices – $b_{i,j}$.

The square shape of the shift matrices is chosen due to the fact that when using a rectangular raster, it is easy to organize the shift without losing information. For the same reason, you can use a hexagonal raster. However, the shape of the matrices can be arbitrary. It is sufficient that the elements of the input image are completely placed in them, and the registers of the cyclic shift were rotated relative to the registers of adjacent matrices by an angle $\Delta\varphi$.

FIGURE 6.3
The relative position of the matrices BM and ShM_φ, for which the positions of the images coincide.

FIGURE 6.4
Copy image elements to shift matrices.

The algorithm for forming the *FAI* for each direction φ is as follows.

1. Image elements from the basic matrix are copied to n corresponding shift matrices (Figure 6.4).
2. Matrix elements must be synchronized. That is, the location of the image elements at the initial moment ($t_{sh} = 0$) for each shift matrix was considered the initial ($b0_{i,j}$).
3. In each matrix ShM_φ there is a simultaneous shift of all available image elements in the directions of the location of the corresponding cyclic shift registers.

4. For each shift moment $(t_{sh} \neq 0)$, if $a_{i + (bsh_{i,j} - b0_{i,j}) \cdot \cos(\varphi), j + (bsh_{i,j} - b0_{i,j}) \cdot \sin(\varphi)} \neq 0$, for this direction φ value $FAI(bsh-b0)$ increases by 1. That is, the intersection of the copy of the image with the original in this element is detected.

5. *FAI* for each direction of shift φ at each moment of time (t_{sh}) is defined as the sum of all elements with non-zero intersection of the original image from the basic matrix and the elements of the corresponding shift matrix.

The block diagram of a possible device for obtaining a set of the functions of the area of intersection is shown in Figure 6.5. The corresponding elements of the matrices are combined by the elements of logical multiplication $(a_{k,m}$ & $bsh_{i,j})$,
where

$$k = i + \left(bsh_{i,j} - b0_{i,j}\right) \cdot \cos(\varphi), m = j + \left(bsh_{i,j} - b0_{i,j}\right) \cdot \sin(\varphi).$$

The zero direction of the image copy offset ($\varphi = 0°$) can be considered in any of the orthogonal directions. With the help of the part of the reference surface created by the data of any two adjacent quarters of the Cartesian coordinate system, it is possible to restore the entire reference surface, which reflects the original figure.

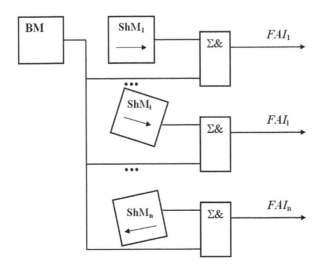

FIGURE 6.5
Block diagram of the device for obtaining a set of *FAI*s.

In the middle of each of these square matrices, we can select a circle-shaped area that is common to all matrices. The image should be located in it initially. Then point 4 of the earlier mentioned algorithm in the device can be organized by directly combining the logical element AND the corresponding elements $a_{i,j}$ and $b_{i,j}$, which form these circular area. The sum of the outputs of all relevant elements of the AND is the value of the *FAI* for a given direction of the shift at a given time.

The initial synchronization of images can be organized by placing them in the center of the matrices. This means that the point of intersection of the diagonals of each of the matrices coincides with the point of intersection of the diagonals of the circumscribed rectangle (CR_0) for the shift direction $\varphi = 0°$. The circumscribed rectangle for the shift direction φ is a rectangle tangent to the sides of the depicted figure, the sides of which are inclined to the orthogonal directions by the angle φ (Figure 6.6).

Similarly, we can organize the creation of a set of *FAIs* using software. However, it should be borne in mind that when rotating the digitized image in certain cells, there will be no exact coincidence of the elements of the basic matrix and the shift matrices. There is a need to correct inaccuracies. This can be done by applying certain intersectional thresholds or determining the presence of an intersection by analyzing the cell environment.

The advantages of parallel formation of functions of the area of intersection for different directions are the speed of this process. However, as the number of such directions (n) increases, hardware costs increase significantly. In addition, it is not possible to form *FAIs* for shift directions that are not included in the selected list.

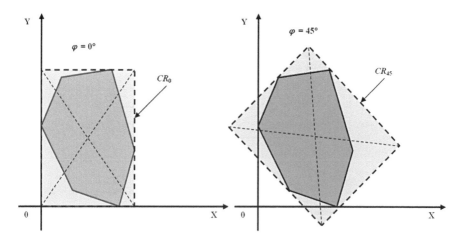

FIGURE 6.6
Examples of circumscribed rectangles for angles of inclinations 0° and 45°.

6.3 Vector Formation of Functions of the Area of Intersection

It is possible to form an *FAI* for any direction of the shift by organizing the offset of the copy of the image in only two directions. Consider the shift of the copy of the image in two orthogonal directions (Figure 6.7).

Any mutual intersection of the original figure from one of the copies or the intersection of the copies with each other corresponds to some value of the *FAI* for a particular direction. For this example, the offset of the image copy in both orthogonal directions is the same ($\Delta x = \Delta y$). Then the mutual intersection of the copies of the image will be equal to the value of the function of the intersection area of the original figure and its copy, which is shifted in the direction $\varphi = 135°$. The zero direction of the image copy offset ($\varphi = 0°$) will be considered the upward direction, which coincides with the direction of the ordinate axis (OY). The offset distance of the image copy is equal to $\Delta\phi = \sqrt{\Delta x^2 + \Delta y^2}$.

The shift of the image elements in a certain direction is determined by the distance and direction. These data can be presented in vector form (Figure 6.8).

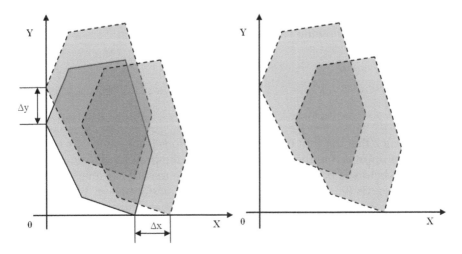

FIGURE 6.7
Shift copies of the figure in two orthogonal directions.

FIGURE 6.8
Formation of the shift vector for the direction $\varphi = 135°$.

Taking into account the property of the central symmetry of set of the *FAIs* when constructing the reference surface, we can determine that the *FAI* value for the direction $\varphi_1 = -45°$ and a similar shift distance will have the same value.

$$FAI(\bar{\varphi}) = FAI(-\bar{\varphi}) \tag{6.1}$$

With a constant shift rate of the copies of the figure in both orthogonal directions, given the property of the central symmetry of the reference surface, and the availability of options for finding the intersection figure–copy or copy–copy, you can get the FAI value only for a limited number of directions. For the Cartesian coordinate system, these are the directions of shift φ, which are equal to $-135°$, $-90°$, $-45°$, $0°$, $45°$, $90°$, $135°$, $180°$. To obtain the value of the functions of the area of intersection for other directions of the image copy shift, it is necessary to modify the shift vectors $\bar{x}$ and $\bar{y}$. To do this, the system for forming a set of *FAIs* must be able to perform image displacement in orthogonal directions at different speeds.

We introduce the use of two parameters $\overline{v_x}$ and $\overline{v_y}$. These are the velocity of the image copy shift in orthogonal directions.

Then, if the corresponding shift time intervals are t_x and t_y, then the shift vectors have the following values.

$$\bar{x} = \overline{v_x} \cdot t_x \tag{6.2}$$

$$\bar{y} = \overline{v_y} \cdot t_y \tag{6.3}$$

We assume that the shift time for both directions is the same ($t_x = t_y$).

Another approach to the implementation of the process of forming modified shift vectors is possible. If it is not possible to shift the image copies at different speeds ($\overline{v_x} = \overline{v_y}$), then the shift can be performed at different time intervals ($t_x \neq t_y$). In both cases, the goal is to obtain vectors with different modules (Figure 6.9a).

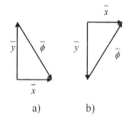

a)   b)

FIGURE 6.9
Formation of the shift vector at different values of orthogonal vectors.

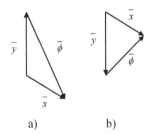

a)   b)

FIGURE 6.10
Examples of shift vector formation at different values of vectors $\bar{x}$ and $\bar{y}$ for hexagonal coverage.

When calculating the values of the functions of the area of intersection for shift directions from 0° to 90°, it is necessary to provide the possibility of shifting the copy of the image vertically not only upwards but also downwards (Figure 6.9b). Or at cyclic shift of the image to define intersection on a vertical in the range from $T_y - Y_{max}$ to T_y, where T_y – the size of a shift matrix in the y direction (in this case on a vertical).

If one of the orthogonal shift vectors of the image copy ($\bar{x}$ or $\bar{y}$) is equal to zero, then the processes of formation of the functions of the area of intersection for orthogonal directions coincide with the processes described in previous works.

If $\bar{y} = 0$, then $FAI(\phi) = FAI(x)$, $\bar{\phi} = \bar{x}$. Parallel shifts of copies of elements of a set occur only in the horizontal direction.

If $\bar{x} = 0$, then $FAI(\phi) = FAI(y)$, $\bar{\phi} = -\bar{y}$. Parallel shifts of copies of elements of a set occur only in the vertical direction.

The process of forming the shift vector for the hexagonal coverage is similar. The only difference is that the angle between the vectors $\bar{x}$ and $\bar{y}$ is not 90°, but 120° (Figure 6.10).

For hexagonal coverage, it is possible to organize a shift without losing information in six directions. For a rectangular coverege, it is possible to organize a shift without losing information in four directions.

For spaces with different dimensions, the processes of formation of the resulting shift vector ($\bar{\phi}$) will be as follows.

The straight line (dimension 1) has two possible shift directions $\bar{x}$ and $-\bar{x}$ ($\bar{y} = 0$). However, it should be remembered the property of central symmetry that is reflected in formula 6.1. Then, when manipulating the elements of a straight line (in a certain way a string of characters can be considered as such of set), it is enough to shift one copy of the object along a straight line to form a function that reflects the interaction of it and its copies.

To determine the vector $\bar{\phi}$ on the plane, it is necessary to organize a simultaneous parallel shift in two directions as shown earlier for rectangular and hexagonal coverage.

Manipulations in three-dimensional space require the presence of three shift vectors of the image copy.

Apart from the plane, there are no geometric objects that are completely covered by elements of the same shape (like hexagons with a hexagonal coverage) other than rectangle. Then, for spaces with dimensions larger than two, rectangular covering objects should be used.

Therefore, to form an arbitrary shift vector of the set in d-dimensional space, it is necessary to be able to organize an orthogonal shift for d directions.

The advantages of this formation of a set of the functions of the area of intersection is that when organizing shifts in orthogonal directions (rectangular coverage) or with an angle of 120° between the shift directions (hexagonal coverage), this process occurs without loss of information. This is the main advantage of this method of forming shift vectors. In addition, it is possible to obtain *FAIs* for any possible direction of displacement. Hardware costs in vision systems are also reduced. The formation of *FAIs* for all possible directions can be organized with only two shift matrices.

The disadvantage of such formation of reference surfaces is the low speed. Although, if it is possible to use two shift matrices for each possible direction, the speed will be maintained.

6.4 Formation of FAIs Sets for Non-Binary Images

The functioning of vision systems using PST is very similar to the functioning of the vision systems of biological beings. The processes of pretreatment and research of plane binary images correlate in some way with the functioning of eye cells such as rods. In this case, the shape of the object is examined. However, for a comprehensive perception of the surrounding world, the IS must be able to operate with data from non-binary images. A lot of information is contained in the color and brightness of their elements.

Consider the reasons for the perception of surface points of three-dimensional objects of research (ORe) as elements of the image different in color and brightness. We will also investigate the interaction of automatic vision devices, light sources, and ORe surfaces.

Among the main questions that an IS must answer, there are the following three. This is the question of "What, Where and When". To interact with the objects of the surrounding world, we need to know which of them are around the IS. Thus, finding the answer to the question "What" refers to solving the problems of object classification. This process is made possible by the correct interpretation of the data obtained by the video system.

The answer to the question "Where" allows you to determine the location of objects in space relative to the IS (for mobile autonomous robots or cars) or relative to the elements of the landscape (for stationary systems). In the case of moving devices, you can take advantage of the properties of binocular vision by placing several video sensors on their body. Stationary systems allow the use of a large number of video cameras, which can be located at considerable distances from each other. If the object of study falls within the field of view of any pair of cameras, its location can be determined by triangulation.

The answer to the question "When" allows you to determine the conditions that affect the time of the visual process. Typically, video surveillance systems use three types of separation of time observation: continuous observation, observation in a certain time range, and observation of the presence of movement in the field of view of the video camera. The required amount of system memory depends on the selection of the desired variant when using these options at save information.

The use of parallel shift techniques in machine vision systems can increase their productivity in the classification of research objects. For the correct interpretation of visual information, it is necessary to understand the processes of formation of this data. The nature of the resulting image is influenced by the type of lighting, the relative position of visual detectors, and objects of research.

The luminous flux detector perceives light that is reflected from the surface of the object of research. The illumination of the object of research can be spot or diffused.

At spot lighting, the light source can be assumed to be smaller in size than the luminous flux detector and located at a considerable distance from it. They can be considered like points. Then the amount of light perceived by the detector from a certain area will be proportional to the projection of the beam reflected from this area towards the detector (Figure 6.11).

To obtain this value, it is necessary to determine the plane at three points: light source (p1), surface area (p2), and light flux detector (p3). In this plane, a line belonging to the surface of the object must be drawn through the point p2.

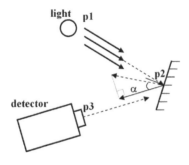

FIGURE 6.11
The magnitude of the luminous flux perceived by the light detector.

The luminous flux obtained by the detector will be proportional to the light intensity of the reflected beam multiplied by the cosine of the angle to the direction of the detector (α).

To simplify the manipulation of the color image in the future, we will consider it as an image converted into a grayscale according to the brightness of the areas. The surface of three-dimensional objects can have both monochrome and multicolor coloration, which can be displayed in one or more shades of gray, respectively. The shape of the surface of the object can consist of plane or convex areas.

From a plane surface, beams of light are reflected in parallel. The flux of light reflected from a three-dimensional surface varies depending on the shape of the surface of the object of research (Figure 6.12).

Due to the large distance between the light detector and the object of research, the beams of light reflected from the plane areas of the surface can be considered parallel.

The effect of misinterpreting plane images when they are visually similar to three-dimensional shapes is used in some optical illusions. The difference

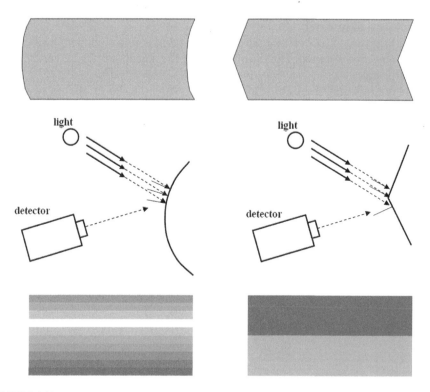

FIGURE 6.12
The perception of the light by the flux detector from objects with different surface shapes.

between a change in color and a change in brightness can only be determined by a change in perspective. If the brightness of certain parts of the object of research becomes brighter in this case, it signals a change in brightness. If the ORe illuminance is constant (diffused light is present), it means that the change in brightness signals a change in the color of the object. If both factors are present, then there is an increase in the brightness of the object, and the difference between the brightness of the colored areas is constant (Figure 6.13).

For manipulations with images of different brightness in machine vision systems, it is necessary to provide the possibility of forming functions of the area of intersection for non-binary images. The processing of colored objects requires the creation of devices that would simulate the work of human visual cells, such as cones.

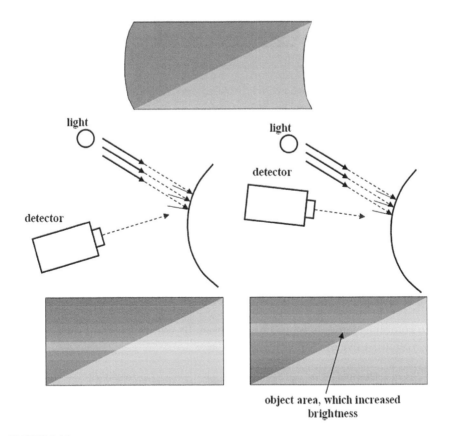

FIGURE 6.13
Perception of light flux from the surface of a two-color object in case of change of position of the detector.

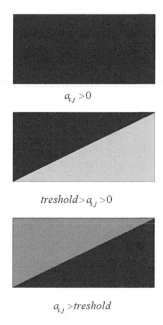

$$a_{i,j} > 0$$

$$treshold > a_{i,j} > 0$$

$$a_{i,j} > treshold$$

FIGURE 6.14
Selection of images with different brightness ranges.

When forming functions of the area of intersection for non-binary images, it must be possible to convert the images. Input images for the visual system can display the entire range of values of non-zero elements or a range of certain values of the brightness of the image elements (Figure 6.14).

The algorithm for forming the functions of the area of intersection is modified as follows. Point 4 reflects the brightness of the relevant elements of the object of research and its copy, which is shifted in parallel.

If $a_{i + (bsh_{i,j} - b0_{i,j}) \cdot \cos(\varphi)' \, j + (bsh_{i,j} - b0_{i,j}) \cdot \sin(\varphi)} \neq 0$, then the image is considered binary.

When $a_{i + (bsh_{i,j} - b0_{i,j}) \cdot \cos(\varphi)' \, j + (bsh_{i,j} - b0_{i,j}) \cdot \sin(\varphi)} = b0_{i,j}$, then the image analyzed is a set of elements whose brightness is equal to the brightness of the element $b0_{i,j}$.

To combine elements whose brightness falls within a certain range of values, you must use the following formula.

$$treshold1 \geq a_{i + (bsh_{i,j} - b0_{i,j}) \cdot \cos(\varphi)' \, j + (bsh_{i,j} - b0_{i,j}) \cdot \sin(\varphi)} \geq treshold2 \qquad (6.4)$$

The values of the variables *threshold1* and *threshold2* can be any. They depend on the type of tasks that need to be solved by the IS.

Changes are also made to the device for obtaining a set of *FAIs* for the range of directions (Figure 6.15).

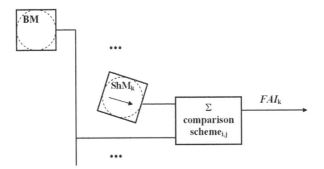

FIGURE 6.15
Block diagram of the device for obtaining FAI_k for a certain shift moment.

In this case, the elements of both the original image and its copies may have values other than 0 and 1. Therefore, the combination of elements of the initial matrix BM ($a_{i,j}$) and the shift matrix ShM_k ($b_{i,j}$), which forms a common circular area (in Figure 6.15 are marked by a dotted line), occurs not with the help of logical elements of AND, but with the help of certain comparison schemes. In these schemes, the elements $a_{i,j}$ and $b_{i,j}$ are compared in accordance with the specified threshold values. To prevent loss of information when organizing shifts, each shift matrix (ShM) can be replaced by two orthogonal shift matrices, where different data shift rates are implemented.

When analyzing the values of the input image elements, it can be divided into several brightness ranges. Then for each of the ranges you need to save your reference surface. Accordingly, the number of devices for obtaining the values of the set of *FAIs* should be the same. In the memory of the IS, each image is stored as a set of reference surfaces for each of the brightness ranges.

6.5 Noise Control with the Use of PST

Parallel image shift can be used in the noise control process. Previous publications on this topic (Bilan & Yuzhakov, 2018; Yuzhakov & Bilan, 2019) proposed a method of noise control by removing contour points. These elements are separated from the original image by searching for information that is not included in the intersection of the object and its copy due to several shifts in different directions.

Figure 6.16 shows the process of determining contour sections by shifting a copy of the image in four orthogonal directions (right, left, top, and bottom). Contour areas are highlighted in black. These are elements that belong to the original image (*In_image*) and did not intersect with its copy (*Copy_image_i*).

FIGURE 6.16
Defining contour sections by four shifts of the image copy in orthogonal directions.

The contour of the image consists of all the contour elements that are defined for each of the shifts.

$$Contour = \sum_{i=1}^{nsh} \left(In\_image \setminus Copy\_image_i \right) \tag{6.5}$$

where *nsh* is the number of shifts of the image copy in different directions.

The accuracy of contour element determination increases with increasing shift directions of the image copy. The width of the contour depends on the value of the offset in each direction. The best contouring accuracy is achieved by shifting by no more than three units (pixels when processing digital images).

Let's decide on the wording. In various sources, images that consist of thin lines are called contour or dashed lines. An example is the image of papillary lines. The basis for constructing the functions of the area of intersection is the processing of figures in which it is easy to determine the area. Accordingly, such figures as shown in Figure 6.2 will be called non-contour.

Since the PST operates mainly with parameters of non-contour images, then we will consider as such noise elements small in size image areas that differ in brightness from the initial image. These can be small groups of pixels or thin lines. For a binary image, the most complex noise of this type will be the inversion of a certain part of its elements (Figure 6.17).

The number of noise elements will be calculated using the percentage of noise (*pn*). This parameter is equal to the percentage of noise elements in the image. Its relationship to the well-known engineering term *PSNR* is shown in the following formula.

$$PSNR = 20 \log_{10} \left(\frac{100}{pn} \right) \tag{6.6}$$

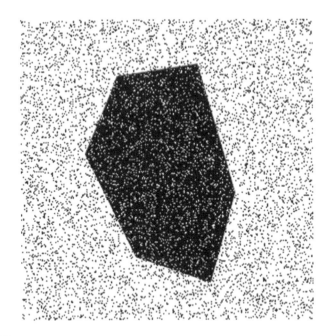

FIGURE 6.17
Inversion of binary image elements ($pn = 10\%$).

The main action in the process of noise control by means of PST is the removal of contour points. Small objects are identified as contours and deleted. Call this action "deleting" ("D"). This process is similar to the "erosion" process described in digital image processing (Gonzalez, Woods, & Eddins, 2004). However, as a result of its implementation, the contour elements of the image are removed, as well as the voids in the middle of the figure increase. This distorts its shape. Elements of the initial image without noise elements are called "useful".

There is a need to compensate for the action of "deleting". For this purpose in the previous works it is offered to combine removal of contour elements with "inversion" ("I"). As a result of this action, the binary image is inverted. The background becomes an image. The contour points are determined and deleted again. Then there is an "inversion". As a result of the combination of "DIDI" actions, the object of research acquires its original shape and size, and a certain amount of noise elements is removed.

We can also restore the shape of the original image with a combination of actions "IDID". A useful consequence of performing this sequence will be the removal of noise zero elements inside the figure. However, the amount of noise in the background areas increases.

To eliminate the negative effects, it is necessary to use a combination of the two specified sequences of actions. The control of noise inclusions in

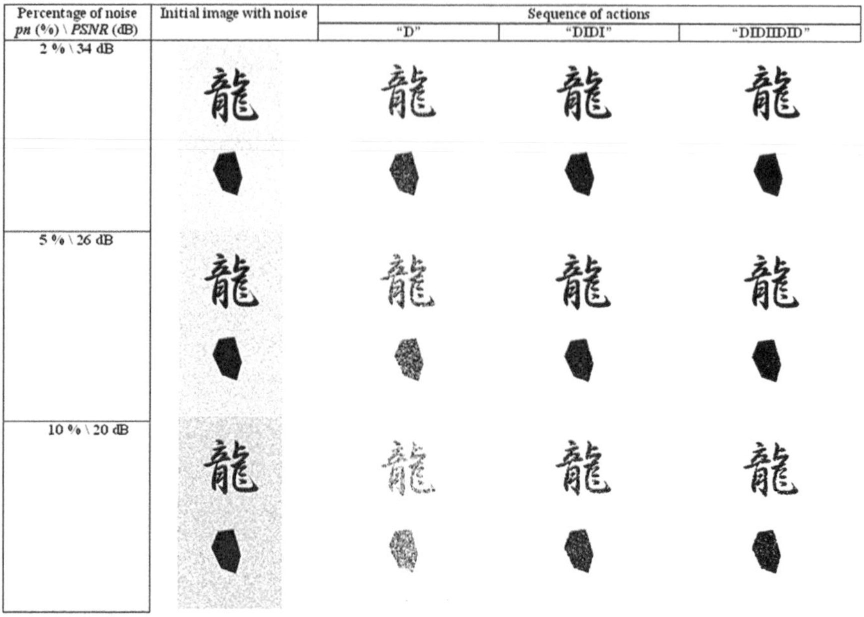

Percentage of noise pn (%) \ *PSNR* (dB)	Initial image with noise	Sequence of actions		
		"D"	"DIDI"	"DIDIIDID"
2 % \ 34 dB	龍 ◀	龍 ◀	龍 ◀	龍 ◀
5 % \ 26 dB	龍 ◀	龍 ◀	龍 ◀	龍 ◀
10 % \ 20 dB	龍 ◀	龍 ◀	龍 ◀	龍 ◀

FIGURE 6.18
Examples of noise control with different combinations of actions at different noise levels.

the pre-processing of binary images is most effective using the sequence "DIDIIDID" (Figure 6.18).

In these examples, the width of the contour to be removed at all stages of noise control was equal to one pixel. Using a wider contour significantly distorts the image. The use of this method at $pn > 10\%$ ($PSNR < 20$ dB) is not effective. Loss of useful information eliminates the effect of noise removal.

We use the indicator "contourness". Contourness is the ratio of the number of contour points of the image to its initial area.

$$Contourness = \frac{Contour}{S_0} \qquad (6.7)$$

The value of this parameter is very important for the preliminary assessment of the effectiveness of the use of PST methods in image processing. The values of contour (*Contour*) and initial area (S_0) are taken for a noise-free image. For contour (dash) images *Contourness* → 1, because they have all (or almost all) elements belonging to the contour. For other images, the contour value is in the range from 0 to 1.

The value of this indicator for non-contour images is approximately equal to the following.

$$Contourness \approx \frac{P}{S_0}. \qquad (6.8)$$

where P is the perimeter of the figure.

Since the image can consist of several parts, then

$$P = \sum_i P_i, \tag{6.9}$$

where P_i is the perimeter of each of the individual parts.

The ratio of equation (6.8) is approximate, because the processes of real information processing make their adjustments. In some figures, certain parameters can be calculated analytically. For example, there is an orthogonally arranged square of 100 × 100 pixels. Side of the circumscribed rectangle (CR_0) $a = 100$. Then analytically calculated value $Contourness \approx \dfrac{4a}{a^2} = \dfrac{4}{a} = 0.04$. Real value $Contourness = \dfrac{396}{10000} = 0.0396$.

The larger the size of the figure, the closer the values of analytically calculated and real parameters.

The real values of the contour parameters for the hieroglyph "dragon" and the test hexagon (Figure 6.18) will be as follows.

$$Contourness_{dragon} \approx 0.154$$

$$Contourness_{hexagon} \approx 0.035$$

Determine the proportion of "useful" pixels of the image relative to the initial number of elements (S_0) that will remain present in the frame after removing noise of different levels due to different sequences of actions for these two images (Tables 6.1 and 6.2). To obtain the earlier mentioned data for each of the two images for each pn for each sequence, 100 attempts were made to perform the noise control algorithm. Testing was performed on software that was created to model these processes.

These data confirm the increase in the efficiency of the noise control process by complicating combinations of simple actions (inversion and removal of contour areas). These processes are more efficient for objects with smaller

TABLE 6.1

The part of "useful" image elements after removing noise from the image of the hieroglyph "dragon"

	D	DIDI	DIDIIDID
pn = 2%	0.67	0.94	0.97
pn = 5%	0.54	0.87	0.94
pn = 10%	0.41	0.72	0.83

TABLE 6.2

The part of "useful" image elements after removing noise from the image of the test hexagon

	D	DIDI	DIDIIDID
pn = 2%	0.76	0.96	0.99
pn = 5%	0.62	0.90	0.98
pn = 10%	0.45	0.76	0.90

contourness. Contour (dash) images of such a noise control system will be perceived as noise.

In the case of processing multi-tone (color) images, they must be converted into a set of binary images using the selected brightness thresholds. In each image from such a set there is a process of noise control. The data brightness elements are then given the initial brightness value. The resulting images are superimposed layer by layer on top of each other. As a result of such actions, noise elements that have not been removed in some layers can be removed in others. The total number of noise elements will decrease.

Qualitative formation of reference surfaces and noise control are important methods of PST at image processing. Pre-processing is an important step in preserving images of research objects and their subsequent classification.

7

Image Recovery by Methods of Parallel Shift Technology

7.1 The Method of Circumscribed Rectangles

The development of most machine vision systems is carried out in the direction of building systems in which the processes are similar to the vision of living beings. Pre-processing involves mechanisms to create information that will be stored in the memory of the IS. It is also desirable to be able to convert the saved data to its original form. The presence of such mechanisms can be useful for modeling encoding–decoding processes.

Digital information processing involves a certain set of attributes of each element of the original image. Such attributes are location coordinates, brightness, and color. In the Cartesian coordinate system, the location of parts of the image can be embedded in the structure of the matrix, which describes the object of research. That is, the elements of the set (pixels) are arranged in accordance with the values obtained by discretization of the plane image. Brightness or color in this case are the values of these elements.

The parallel shift technology involves operating not with sets of ordered individual elements of the image, but with sets of functions obtained in a certain way. For a binary image, these may be the functions of the area of intersection. Their formation requires the use of simple technical actions. They are determining the area of the object, organizing the displacement of a copy of the image, and detecting the intersection of two sets.

Due to the generalization of the input information (elements of the set are collapsed into a function), the process of restoring the image to its original state is quite complex. The paper (Yuzhakov, 2019) shows the process of restoring convex figures using a search among a number of options.

The basis of any data recovery is to find the area of their location. For example, in the process of analyzing a function, the first step is to evaluate its area of definition. The elements of each image are located within a specific area. For a convex figure, this area coincides with its contour. The elements of a non-convex image are limited by its contours, which in turn are part of some convex area.

To determine the convex area of the image elements, we use the method of circumscribed rectangles. The circumscribed rectangle (CR_φ) is a rectangle tangent to the sides of the image (Figure 6.6). Its sides are inclined relative to the orthogonal directions by a certain angle φ.

The method of circumscribed rectangles is based on the following fact. The shape of the intersection of those CR_φ which correctly located on the plane approaches the shape of the object of study with increasing number of directions φ, for which these rectangles are created (Figure 7.1).

The arrangement of the set CR_φ is true such that the centers of mass of the figures (cm), around which the circumscribed rectangles are constructed, coincide.

Suppose there are circumscribed rectangles for some shape for directions 0 to $\frac{\pi}{2}$. Rectangles for directions from $\frac{\pi}{2}$ to π repeat them (width and height parameters change places); therefore, it is not efficient to use CR for such directions. The parameters of the circumscribed rectangles for the direction i will be considered as follows. A_i is the width of the rectangle, B_i is the height of the rectangle. The value of the intersection area of all available circumscribed rectangles for this figure is called $S_{crossing}$. S_0 is one of the basic parameters of the figure (initial area). Then the statement about the dimensions

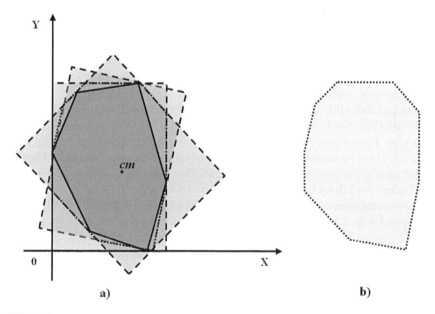

a)   b)

FIGURE 7.1
The intersection of the three circumscribed rectangles (a) and the figure created by this intersection (b).

of the intersection of the circumscribed rectangles can be expressed as the following equation.

$$A_i B_i \cap A_j B_j \geq A_0 B_0 \cap \dots \cap A_i B_i \cap \dots \cap A_{\frac{\pi}{2}} B_{\frac{\pi}{2}} = S_{crossing} \geq S_0 \qquad (7.1)$$

If the number of directions for which circumscribed rectangles are constructed increases, then $S_{crossing} \to S_0$

To correctly position the circumscribed rectangles when restoring the image, we need to know the location of the center of mass of the figure for each rectangle. In the absence of relevant data in the reference databases that describe the object (Bilan, Yuzhakov, & Bilan, 2014b), this process is quite resource-intensive. In this case, it is necessary to perform a sufficiently large sequence of actions to search for possible options for the location of the circumscribed rectangles.

Reducing the number of options can include in the reference databases information about the location of the center of mass for each direction of shift of the image copy. To facilitate the scaling process, the data on the coordinate of the center of mass of the figure should be stored in relative values. The center mass ratio (cmr_φ) indicator must be added to the reference databases. This value is the ratio of the coordinate of the center of mass of the image ($x_{cm\varphi}$) and the maximum shift $X_{max\varphi}$ for the direction φ.

$$cmr_\varphi = \frac{x_{cm\varphi}}{X_{max\varphi}} \qquad (7.2)$$

The value of $X_{max\varphi}$ is the reference. It corresponds to a certain reference value S_0 of a certain object (Figure 7.2) and is calculated from the parameters of the *FAI* at the stage of training of the system (creating a reference surface of this image).

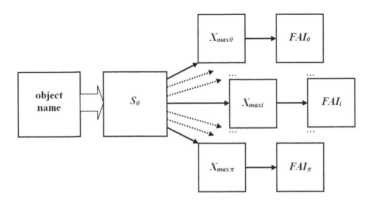

FIGURE 7.2
Saving in reference databases the values of the sequence of parameters for a certain object.

When scaling the values of the standards according to some scaling coefficient (k_{sc}), the new values of the initial area ($S_{0\varphi new}$) and the maximum shift ($X_{max\varphi new}$) will be as follows.

$$S_{0\varphi new} = k_{sc}^2 \cdot S_{0\varphi} \qquad (7.3)$$

$$X_{max\varphi new} = k_{sc} \cdot X_{max\varphi} \qquad (7.4)$$

The value of the coordinate of the center of mass for each direction changes accordingly.

$$x_{cm\varphi new} = cmr_\varphi \cdot X_{max\varphi new} = cmr_\varphi \cdot k_{sc} \cdot X_{max\varphi} \qquad (7.5)$$

The coordinate value of the center of mass ($x_{cm\varphi}$) is determined relative to the left boundary of the circumscribed rectangle for this direction. This value does not need to be stored in the reference databases. It is enough to save the parameter cmr_φ, which does not depend on scaling. However, to calculate this indicator according to formula (7.2) at the training stage (filling in data for a specific standard), it is necessary to obtain the value of this coordinate. To do this, a non-cyclic shift of the image copy is performed in the shift matrix (ShM_φ) and the area of the part of the copy of the image that is in the field of the shift matrix (S_i) is calculated. The shift value is fixed for the moments when the area of the part of the figure begins to decrease (*shift_dec*), when the area of the part of the copy of the figure does become equal to $\dfrac{S_0}{2}$ (*shift_cm*), and when the area of the part of the copy of the figure does become equal to 0 (*shift_0*). Example of calculating the coordinate of the center of mass of the test figure is shown in Figure 7.3.

The value of the width of the circumscribed rectangle for this direction of displacement will be as follows.

$$A_\varphi = shift\_0 - shift\_dec \qquad (7.6)$$

The value of the coordinate of the center of mass of the figure, which is used to calculate the parameter cmr_φ, is obtained from formula (7.7).

$$x_{cm\varphi} = A_\varphi - (shift\_cm - shift\_dec) = shift\_0 - shift\_cm \qquad (7.7)$$

As can be seen from the latter formula, the calculation of the values of the coordinates of the center of mass of the image can be performed regardless of the linear size of the shift matrix (T_φ) and the width of the circumscribed rectangle (A_φ). These parameters are important when processing objects by PST methods and will be used in the future.

The vertical coordinate of the point cm_φ ($y_{cm\varphi}$) for the shift direction φ is equal to the coordinate $x_{cm\left(\varphi+\frac{\pi}{2}\right)}$ for the shift direction $\varphi + \dfrac{\pi}{2}$.

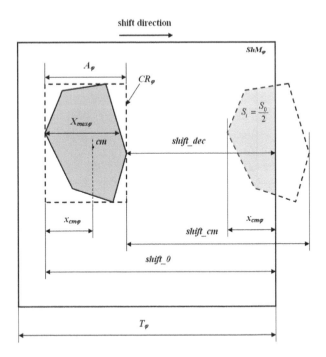

FIGURE 7.3
Example of calculating the coordinate of the center of mass of the test figure.

Assuming that the parameters of the shift matrix for the direction φ are equal to the parameters of the circumscribed rectangle CR_φ, formula (7.6) will have the form $A_\varphi = shift\_0$, and formula (7.7): $x_{cm\varphi} = A_\varphi - shift\_cm$. However, this option is not easy to organize technically because the dimensions of the circumscribed rectangles for different shapes are different. In addition, the connection with the physical location of the image on the plane is lost.

Before combining the coordinates of the centers of mass of the figure for all available circumscribed rectangles, it is necessary to list the parameters of the lines on which their sides lie, to translate into one coordinate system CR_0 (Figure 7.4). To do this, it is sufficient to know the coordinates of the point cm_0, the values of the angles φ, the values of the parameters of the circumscribed rectangles (A_φ and B_φ), and the values of the coordinates of the points cm_φ for each of the directions ($x_{cm\varphi}$ and $y_{cm\varphi}$). They define perpendiculars to the sides of the circumscribed rectangles.

The use of the cmr_φ parameter in the database of descriptions of the object will correct the consequences of the fact that the mirror images have the same *FAIs* for the direction of displacement perpendicular to the axis of symmetry, which negatively affects the image recognition processes.

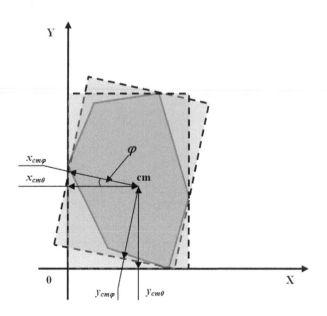

FIGURE 7.4
Combination of the coordinates of the centers of mass of images for two circumscribed rectangles.

7.2 Determining the Parameters of Circumscribed Rectangles

The application of the method of circumscribed rectangles allows to restore a convex image from the data of the reference surface. For non-convex images, only the area of definition of their elements is determined. The values of the sides of the circumscribed rectangles can be determined not only by using the shift parameters of the *ShM* matrices, but also by the values of the functions of the area of intersection that form the reference surfaces.

Criterion of convexity of a figure in case of application of PST: *each following value of FAI of a figure for each direction of shift is less than the previous one.*

$$FAI_\varphi(i+1) < FAI_\varphi(i) \tag{7.8}$$

The intersection of convex images with their copies has the property of equal areas: *the area of the part of the image, and the area of the part of its copy that is shifted in parallel, which are not included in the intersection, are equal* (Figure 7.5).

If the shift $x \to 0$, then the area S_x is close in value to the area of the line that reflects the contour of the figure. Using this property, we can determine the parameters of the circumscribed rectangle for convex shapes.

$$B_0 = FAI_x(0) - FAI_x(1) = S_0 - FAI_x(1) \tag{7.9}$$

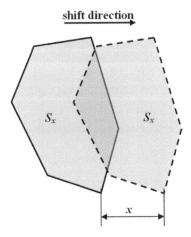

FIGURE 7.5
The property of equality of areas when shifting a copy of the image.

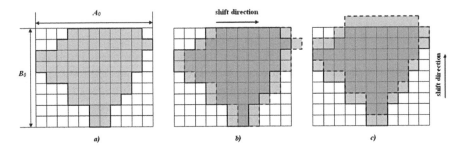

FIGURE 7.6
Initial convex image (a) and shifts of its copies by one unit in orthogonal directions (b, c).

$$A_0 = FAI_y(0) - FAI_y(1) = S_0 - FAI_y(1) \tag{7.10}$$

Consider some convex figure and the intersection with its copy in the case of single orthogonal shifts (Figure 7.6).

$$S_0 = 59$$

$$B_0 = S_0 - FAI_x(1) = 59 - 50 = 9$$

$$A_0 = S_0 - FAI_y(1) = 59 - 48 = 11$$

The width and height of the circumscribed rectangle of the convex figure are equal to the values of the parameters of the parts of its contour with a

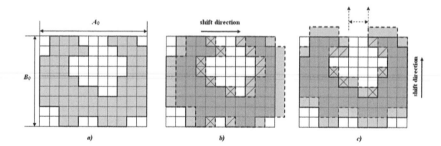

FIGURE 7.7
Initial non-convex image (a) and shifts of its copies by one unit in orthogonal directions (b, c).

width of one pixel. Thus, it is possible to determine the parameters of the circumscribed rectangles of convex images based on the data of the reference surfaces.

Criterion of convexity of a part of a convex image: *if a figure is convex, then any intersection with its copy, or the intersection between its copies, is a convex image.* Then, for each intersectional variant, the parameters of the circumscribed rectangles for a certain shift direction can be calculated on the basis of the reference surface data for this direction.

For non-convex images, it is necessary to make certain adjustments when calculating the parameters of the circumscribed rectangles (Figure 7.7).

Images can be segmented into specific bars that match the shift direction. In the technical implementation, they will correspond to certain parts of the shift registers of the ShM_φ matrices. These circumscribed rectangles are part of shift matrices.

In Figure 7.7, in addition to the gray elements of the original image and its copy which marked by the dotted line, there are elements marked with images (⊠ and ⊿).

These designations signal the following.

⊠ – the element belongs to the copy of the figure and does not belong to the original image (let's call it $copy\_ctr_{ij}$).

⊿ – the element belongs to the original image and does not belong to the copy of the figure (let's call it $image\_ctr_{ij}$).

$$image\_ctr_{ij} = in\_image_{km} \wedge \overline{copy\_image_{ij}} \tag{7.11}$$

$$copy\_ctr_{ij} = \overline{in\_image_{km}} \wedge copy\_image_{ij} \tag{7.12}$$

The parameters k and m are the coordinates of the image elements of the matrix BM, and i and j are the coordinates of its copy on the corresponding shift matrix. The coordinates of the matrices that correspond to the same image elements are denoted by different letters, because the numbers of the elements of the matrices BM and ShM_φ do not match (Figure 6.4). These elements belong to the circular region, which is common to all shift matrices. The matrices are inverted at some angle $\Delta\varphi$ relative to each other.

The same properties are inherent in the outer edge elements of the image and its copies. However, we will not single them out. This will help to understand the reasons for the formation of adjustment parameters.

In the given example (Figure 7.7), shifts of copies of the image are orthogonal. Applying formulas 17 and 18 to this image, we find that when calculating the parameters of the circumscribed rectangle, it is necessary to apply adjustments ($\Delta vert$ when determining B_0 and $\Delta horiz$ when determining A_0).

The real values of the parameters of the circumscribed rectangles can be obtained at the stage of perception (learning) by formula 14 (we mean that $B_0 = A_{\frac{\pi}{2}}$).

Let's calculate the values of the adjustment parameters ($\Delta vert$ and $\Delta horiz$) for the given example ($B_0 = 9$ and $A_0 = 11$).

$$S_0 = 64$$

$$B_0 = S_0 - FAI_x(1) - \Delta vert = 64 - 48 - \Delta vert = 16 - \Delta vert = 9. \ \Delta vert = 7$$

$$A_0 = S_0 - FAI_y(1) - \Delta horiz = 64 - 49 - \Delta horiz = 15 - \Delta horiz = 11. \ \Delta horiz = 4$$

There must be a *copy_ctr$_{ij}$* element in each row of elements of the corresponding ShM_φ matrix, where the *image_ctr$_{ij}$* element is present. These elements are always arranged in pairs. Then the values of the adjustment parameters ($\Delta vert$ and $\Delta horiz$) are equal to the number of elements *image_ctr$_{ij}$* (or *copy_ctr$_{ij}$*) for this direction of shift. Thus, the formulas of the parameters of the circumscribed rectangles receive certain adjustments.

$$B_0 = FAI_x(0) - FAI_x(1) - \Delta vert = S_0 - FAI_x(1) - \Delta vert \qquad (7.13)$$

$$A_0 = FAI_y(0) - FAI_y(1) - \Delta horiz = S_0 - FAI_y(1) - \Delta horiz \qquad (7.14)$$

In this example of determining the parameters of a non-convex shape in each string, there is no more than one pair of elements *image_ctr$_{ij}$* and *copy_ctr$_{ij}$*. Consider an example of forming the parameters of circumscribed rectangles for an image created by several objects (Figure 7.8).

$$S_0 = 43$$

$$B_0 = S_0 - FAI_x(1) - \Delta vert = 43 - 23 - \Delta vert = 20 - \Delta vert = 9. \ \Delta vert = 11$$

$$A_0 = S_0 - FAI_y(1) - \Delta horiz = 43 - 33 - \Delta horiz = 10 - \Delta horiz = 11 \ \ \Delta horiz = -1$$

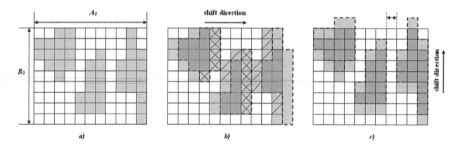

FIGURE 7.8
Initial image which is created by several objects (a) and shifts of its copies by one unit in orthogonal directions (b, c).

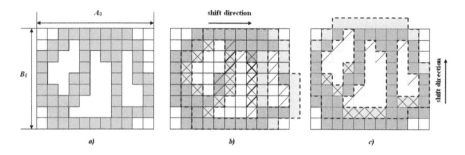

FIGURE 7.9
The initial image of a convex figure with voids (a) and shifts of its copies by one unit in orthogonal directions (b, c).

The calculation of the height of the circumscribed rectangle (B_0) in this case corresponds to the values obtained from formula (7.13). The adjustment parameter $\Delta horiz$ is negative. This indicates that between the parts that form the image for the direction of displacement are intervals of total length $|\Delta horiz|$. The presence of *image_ctr$_{ij}$* or *copy_ctr$_{ij}$* elements increases the adjustment parameters for the respective directions, and the presence of gaps between parts of the image reduces them. In other cases, the value of $\Delta vert$ may be negative. In any case, these values can help in the process of segmenting the object of research into individual parts or finding the distance between objects in the analysis of scenes.

The following is an example of determining the parameters of the circumscribed rectangle for a convex image with several voids inside (Figure 7.9).

$$S_0 = 57$$

$$B_0 = S_0 - FAI_x(1) - \Delta vert = 57 - 32 - \Delta vert = 25 - \Delta vert = 9. \ \Delta vert = 16$$

$$A_0 = S_0 - FAI_y(1) - \Delta horiz = 57 - 37 - \Delta horiz = 20 - \Delta horiz = 11. \ \Delta horiz = 9$$

This example is characterized by the fact that some elements (Figure 7.9b) are both *image_ctr_{ij}* elements and *copy_ctr_{ij}* elements. This situation is caused by the presence of fine lines perpendicular to the direction of shift.

If at the stage of perception of the visual image (or learning of the system), the values of real circumscribed rectangles are not fixed, then in formulas (7.13) and (7.14) there will be two pairs of unknown quantities (B_φ and $\Delta vert$, A_φ and $\Delta horiz$). Then, to calculate the parameters of the circumscribed rectangles, the adjustment parameters based on the data of the reference surfaces can be selected by the method of searching from several options. They can take values in the following ranges.

$\Delta vert$ in the range from $-(S_0 - FAI_x(1) - 2)$ to S_0.
$\Delta horiz$ in the range from $-(S_0 - FAI_y(1) - 2)$ to S_0.

Negative values of ranges occur if there are gaps between the elements of the image, which consists of several objects (similar to the example in Figure 7.10c). The two in the negative part of the range exists because the number of lines that form the image and which are located along the direction of the shift is at least two (Figure 7.10).

For convex shapes, the values of the adjustment parameters are equal 0.

As you can see, even the analysis of the values of the parameters of the circumscribed rectangles (B_φ and A_φ) and the parameters of their adjustment ($\Delta vert$ and $\Delta horiz$) can give some information about the object of research. Knowledge of these values will help in applying the method of circumscribed rectangles.

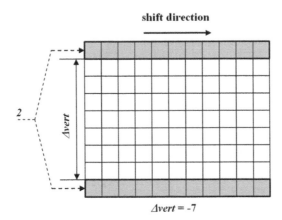

FIGURE 7.10
The reasons for the formation of the minimum value of the adjustment parameter.

7.3 Formation of Reference Surfaces for Parts of the Existing Standard

Determining the parameters of the circumscribed rectangles is possible both with the help of the receptors of the intelligent system at the stage of perception of visual information and with the help of the parameters of the reference surface. If the image is quite complex (non-convex or consists of several objects), then it is necessary to find the parameters of the circumscribed rectangles for certain parts of it. Since the parameters of the circumscribed rectangles can be determined using the parameters of the reference surface, it is necessary to construct reference surfaces for parts of the image, using the available reference surface of the whole figure.

The reference surface is a set of functions of the area of intersection of a certain object for the directions from φ to $\varphi + \pi$, where φ is the direction of the initial scanning of the image (Bilan, Yuzhakov, & Bilan, 2014b). The reference surfaces have the property of central symmetry, so it is not necessary to store the *FAIs* for directions from π to 2π.

The principles of formation of reference surfaces for convex and non-convex images are the same. Therefore, for the simplicity of the study of the process of construction of reference surfaces for part of the image, it will be carried out on the basis of the study of a convex figure. Consider some image A (Figure 7.11a). Figure 7.11b shows the offset of its copy by a distance x_i. As a result, a figure of intersection B is formed (Figure 7.11c), which is also convex.

Figure B is part of the original figure A. Obviously, the intersection of figure B with its copy for each moment of the shift in the direction of shift φ will

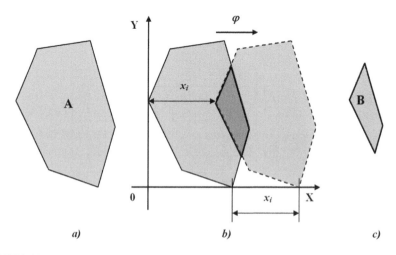

FIGURE 7.11
An example of forming a intersectional figure B when shifting the original image A.

be equal to the value of the *FAI* of figure A for the same direction in the range from x_i to X_{max}. In this example, the direction φ coincides with the abscissa. This direction will be considered zero. It is necessary to obtain a set of functions of the area of intersection of figure B for other directions to construct the reference surface of this image.

The construction of the functions of the area of intersection of a certain acrossing area for figure B for directions of shift, which are not explicitly present, must be performed using the criterion of equality of areas of identical images: *identical images have the same area*.

Consider the value of the intersection of figure B and its copy when shifted by x_{Bj} in the direction φ_{Bj} (Figure 7.12).

The values in this figure mean the following.

x_i is the shift distance of the copy of the image A, which resulted in the shape of the intersection figure B.

x_{Bj} is the shift distance of the copy of the intersection figure B in the direction φ_{Bj}, which resulted in the value of the *FAI* of the reference surface of the image B for this shift in this direction.

$x_{A\alpha}$ is the distance of the shift of copy of the image A in the direction α (the value of the intersectional area is obtained from the reference surface of the original image).

According to the criterion of equality of areas of identical images, the value of the intersection area of figure A with its copy when shifted by the distance $x_{A\alpha}$ in the direction α is equal to the value of the intersection area of figure B with its copy when shifted by distance x_{Bj} in the direction φ_{Bj}, because the intersection figures are the same.

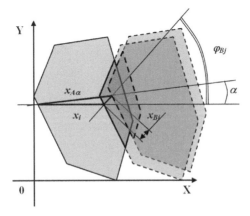

FIGURE 7.12
An example of the application of the criterion of equality of areas of identical images.

The data of the intersection parameters of figure A with its copies for differ-ent shift directions are stored in the reference surface. The data of the param-eters of the intersection of figure B with its copies for different directions of the shift must be determined. The relationship between the shift parameters of the copies of figure A and its part B for arbitrary directions is given in for-mulas (7.15) and (7.16).

$$x_{A\alpha} = \left|\overline{\varphi_{Bj}}\right| = \sqrt{\left(x_i + x_{Bj} \cdot \cos\left(\varphi_{Bj}\right)\right)^2 + \left(x_{Bj} \cdot \sin\left(\varphi_{Bj}\right)\right)^2}$$
$$= \sqrt{x_i^2 + 2 \cdot x_i \cdot x_{Bj} \cdot \cos\left(\varphi_{Bj}\right) + x_{Bj}^2} \qquad (7.15)$$

$$tg\left(\alpha\right) = \frac{x_{Bj} \cdot \sin\left(\varphi_{Bj}\right)}{x_i + x_{Bj} \cdot \cos\left(\varphi_{Bj}\right)} \qquad (7.16)$$

Figure 7.13a schematically shows an example of the reference surface of the initial figure A (top view). The arrows indicate certain directions of FAI formation. The length of each arrow is equal to the magnitude of the maxi-mum shift in this direction. At point 0, the values of all FAIs are equal to S_0. The dotted lines indicate the directions of the functions of the area of

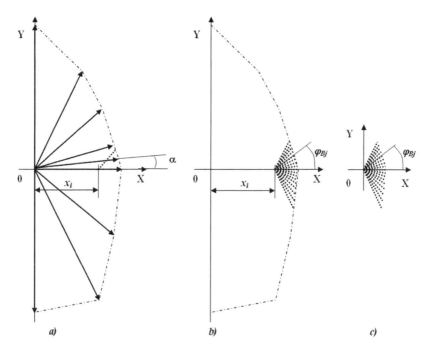

FIGURE 7.13
The formation of the reference surface of a part of the original figure.

intersection for figure B, which can be determined. The values of the *FAIs* for figure B at these points coincide with the corresponding values of the *FAIs* of the original image A (Figure 7.13b).

Since the reference surfaces have the property of central symmetry, the *FAIs* of figure B when shifted in directions opposite to the indicated points are similar. The reference surface of the initial figure is saved for directions from 0 to π. For this example, the vertical direction (up) is selected as zero. The reference surface of a part of the initial image (Figure 7.13c) depending on its form can be constructed for smaller range of values of directions of shift.

Any image has points of contact with the circumscribed rectangles. In the given example, where the initial image is a test hexagon, there are four such points (Figure 7.14). Depending on its shape, the image may touch each side of the CR_φ at one or more points. The maximum value of the number of points of contact may not exceed the dimensions of the corresponding circumscribed rectangle (A_φ and B_φ). Lines that touch the original image pass through the points of contact. A pair of these tangent lines determines the range of formation of directions for the reference surface of the image B.

In the given example for the touch point 1, this range β_1 consists of two angles formed by the tangents at this point (β_{11} and β_{12}).

$$\beta_1 = \beta_{11} + \beta_{12} \tag{7.17}$$

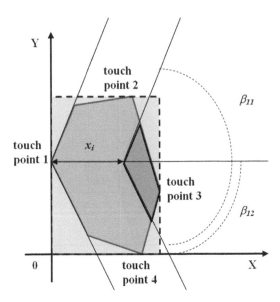

FIGURE 7.14
Determining the range of directions for the formation of the reference surface of a part of the original image.

For touch point 3, at a horizontal shift to the left, a similar range β_3 consists of two angles formed by tangents at this point (β_{31} and β_{32}). Since the pairs of angles (β_{11} and β_{32}, β_{31} and β_{12}) can be different, and the reference surface has the property of central symmetry, the general range of directions of formation of the functions of the area of intersection for the part of the initial image (β) will be as follows.

$$\beta = \max\left(\beta_{11}, \beta_{32}\right) + \max\left(\beta_{31}, \beta_{12}\right) \tag{7.18}$$

The *FAIs* value of the part of the original image (in this case B) in this range can be found in full compliance on the reference surface of the original image (in this case A).

The test hexagon has only one touch point with the sides of the orthogonally located circumscribed rectangle (CR_0). If there are more points of contact to the sides CR_φ, then some adjustment is necessary to determine the range of formation (β) of the set of functions of the area of intersection of a part of the initial figure (Figure 7.15).

For that side of the circumscribed rectangle for which there are several touch points when determining the range of formation of the reference surface of the intersection area, one point of contact is divided into two. In the example in Figure 7.16, the initial figure touches with the left boundary of the circumscribed rectangle on some segment. If there are several touch points, then the points of contact with the minimum (touch point 1_2) and maximum

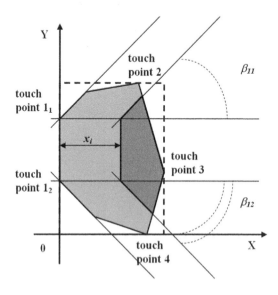

FIGURE 7.15
Determining the range of directions for the formation of the reference surface of the intersection area in the case of several touch points to one side of the circumscribed rectangle.

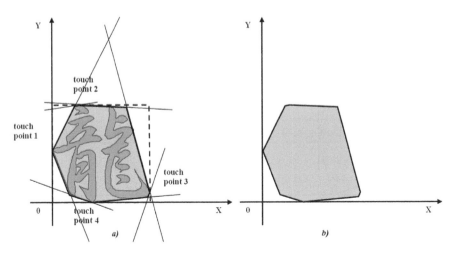

FIGURE 7.16
Formation of the definition area of the elements of the non-convex initial image.

(touch point 1_1) coordinates are selected for the vertices of the corresponding angles β_j (where j is the side number of the circumscribed rectangle). The following steps of forming the reference surface of the intersectional area of the original figure and its copy do not differ from the earlier mentioned ones.

If tangents to the original figure are drawn through the touch points (Figure 7.16a), they limit the definition area for this image (Figure 7.16b).

This process is similar to constructing the intersection of circumscribed rectangles for shift directions that coincide with the tangent directions. Thus, when forming the definition area of the elements of the original image can be limited to a set of circumscribed rectangles, the directions of which are close to the tangent to the image, which pass through the touch points of a single CR_φ (in this example CR_0).

Circumscribed rectangles can be constructed around the initial image and areas of intersection with its copy at each moment of shift (Figure 7.17).

The values of these circumscribed rectangles for convex images will be as follows.

$$A_i = A_0 - x_i = S_0 - FAI_y(1) - x_i \tag{7.19}$$

$$B_i = FAI_x(x_i) - FAI_x(x_i + 1) \tag{7.20}$$

For non-convex images, adjustments are possible based on formulas (7.13) and (7.14).

$$B_i = FAI_x(x_i) - FAI_x(x_i + 1) - \Delta vert \tag{7.21}$$

$$A_i = A_0 - x_i = S_0 - FAI_y(1) - x_i - \Delta horiz \tag{7.22}$$

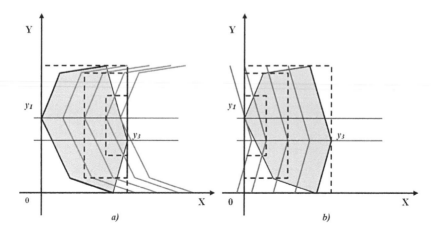

FIGURE 7.17
Circumscribed rectangles around the intersection areas for the horizontal direction of the image copy offset right (a) and left (b).

The coordinates of the touch points which are perpendicular to the direction of displacement in the coordinate system of the circumscribed rectangle are the same for all cases. For a convex image, you can construct all possible touch points using the parameters of the center of mass and the parameters of the circumscribed rectangles for all directions. Their sequential connection will allow us to determine the contour of a convex figure. The parameters of the initial image are enough for this. Filling the area bounded by the contour with elements of the appropriate value ("1" for a binary image) will completely restore the convex image. The accuracy of the recovery depends on the number of shift directions for which the *FAIs* values of the reference surface exist.

For non-convex images, such manipulations with the data of the original figure will only find the definition area of their elements. The reference data of the parts of the original image does not store the values of the parameters of the location of the centers of mass of the image data. The analysis of such images should be carried out with the help of data of their reference surfaces, coordinates of parameters of touch points, and angles of tangent lines at these points. Also, non-convex images can be segmented so that they are a set of convex images. Then you need to restore these convex images, then the sum of their elements will form the original shape.

The feature of initial image recovery is optional. For an intelligent system, it does not matter in what form the information is received, stored, and processed, if at the stage of interpretation of the next possible actions of the system, the correct conclusions are made. The correct conclusions are those that allow for IS to interact with the outside world so as to obtain a neutral or positive value of the "criterion of interaction". Interaction criterion is a parameter that reflects the total efficiency of the system at a certain point in time (the sum of gains and losses).

8

Scene Analysis for Two Objects

8.1 Determining the Basic Parameters of the Scene

One of the main tasks of video information processing is scene analysis. A scene in the visual area will be called as one with the presence of two or more separate objects in it. Scene analysis is divided into two subtasks. The first subtask is to find the relative position of the objects that create the scene. The second subtask is to determine the form and quantitative parameters of each of them.

Consider an elementary scene, which is created by two figures.

From the point of view of application of parallel shift technology, the following options of a mutual placing of two components (A and B) are possible.

1. Objects A and B are located at a distance greater than the linear size of the largest of them in the direction of the shift (Figure 8.1a).

2. Objects A and B are located at a distance that is less than the linear size of the largest of them in the direction of the shift (Figure 8.1b).

3. Objects A and B intersect (Figure 8.1c).

In a scene, objects are characterized by shape, area, and distance between them.

If the sizes of the maximum shifts in the selected direction φ are equal to I_{maxA} and I_{maxB}, respectively, then in the first case the distance between objects ($Dist_{AB1}$) for this direction must be greater than the maximum of them.

$$Dist_{AB1} > \max\left(I_{maxA}, I_{maxB}\right) \tag{8.1}$$

It should be noted that here and subsequently the order of the characters in the variable $Dist$ indicates from which and to which object the distance is determined. In the case of cyclic shift, there are two distances between objects. Indexes 1 to 3 indicate a possible relative position (Figure 8.1).

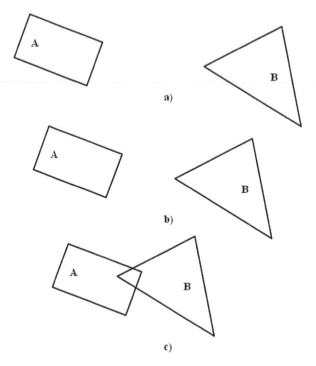

FIGURE 8.1
Options for possible mutual location of two objects.

In the second case, this distance will be as follows.

$$0 \le Dist_{AB2} \le \max\left(I_{\max A}, I_{\max B}\right) \tag{8.2}$$

When objects intersect, the distance between them is less than zero.

$$Dist_{AB3} < 0 \tag{8.3}$$

For application of the methods of parallel shift technology, it will be important how the relative position of the objects will affect their total *FAI*.

Assume that the initial area of the objects is S_{0A} and S_{0B}, respectively, then the total initial area S_{0AB} for cases 1 and 2 of its relative position will be as follows.

$$S_{0AB} = S_{0A} + S_{0B} \tag{8.4}$$

In case 3, it will be reduced by the intersectional area $S_{A \cap B}$.

$$S_{0AB} = S_{0A} + S_{0B} - S_{A \cap B} \tag{8.5}$$

Consideration of variant 3 of the relative position is one of the most difficult tasks in the analysis of scenes. The only thing that can be said with certainty about the total initial area is the following.

$$S_{0A} + S_{0B} \geq S_{0AB} \geq \max\left(S_{0A}, S_{0B}\right) \tag{8.6}$$

Objects A and B can be positioned relative to each other in any way.

$$0 \leq S_{A \cap B} \leq \min\left(S_{0A}, S_{0B}\right) \tag{8.7}$$

When the intersection of the objects is absent, then formula 35 is converted to formula (8.4). That is, formula (8.5) is general.

Thus, using the above mathematical formulas, we can estimate two of the three indicators (initial areas and distances between objects). The third characteristic feature (shape) can be determined only by analyzing the functions of the area of intersection (Belan & Yuzhakov, 2013a; Bilan & Yuzhakov, 2018).

For option 1 of the relative position of the objects, with no intersection between objects and/or their copies, the total function of the area of intersection ($FAI_{AB}(i)$) for the selected direction φ will be as follows.

$$FAI_{AB}(i) = FAI_A(i) + FAI_B(i) \tag{8.8}$$

where i is the shift in the selected direction φ.

For case 2 of the relative position with a small distance between objects when shifting in certain directions, there are certain sectors of the intersection. There are two of them. Each has a range of angle values for possible shift directions equal to α. The intersection sectors limit the shift directions of the image copy for which the intersection is possible. For the other two sectors of size $\pi - \alpha$ FAIs are formed similarly to the previous case (Figure 8.2).

When shifting the copy of the image in the directions that are part of the intersection sectors, formula 38 requires modification.

$$FAI_{AB}(i) = FAI_A(i) + FAI_B(i) - FAI_{A \cap B}(i - Dist_{AB}) \tag{8.9}$$

where i is the offset in the selected direction φ. In this case, the value of $Dist_{AB}$ is equal to $Dist_{AB2}$, $FAI_{A \cap B}(i)$ is the function of the intersection of objects A and B for the selected shift direction φ at their current spatial location.

The functions $FAI_A(i)$ and $FAI_B(i)$ decrease from the values of S_{0A} and S_{0B}, respectively, to 0. They are formed by the intersection of similar figures ($A \cap A$, $B \cap B$). The intersection function $FAI_{A \cap B}(i)$ increases from 0 to a certain maximum value and then decreases again to 0. In this case, different objects ($A \cap B$, $A \neq B$) intersect. The property described in formula 39 is inherent in any objects that are at a small distance from each other.

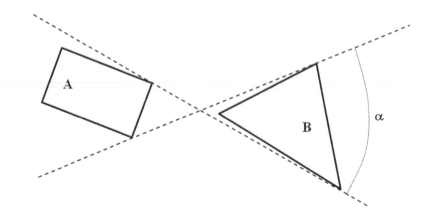

FIGURE 8.2
Presence of intersection sectors.

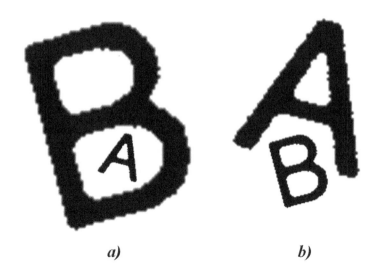

a)   *b)*

FIGURE 8.3
Some options for the location of non-convex objects.

The shape and the relative position of objects may be such that one of them is completely (Figure 8.3a) or partially (Figure 8.3b) inside the other.

A prerequisite for this type of location is that both intersection sectors have a size that is equal to π. In this case, the following inequality is always true.

$$\min\left(S_{0A}, S_{0B}\right) \geq FAI_{A \cap B}\left(i\right) \neq 0 \qquad (8.10)$$

This relation is valid for all possible directions of the image copy shift for those values of i for which there is a mutual intersection. The objects of

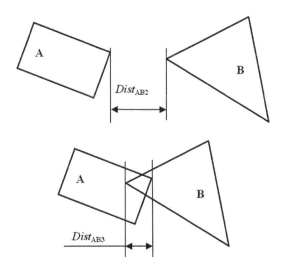

FIGURE 8.4
Formation of distances between objects.

research (A and B) can be of any shape and size, can be turned at any angle relative to orthogonal directions, and can intersect in the process of shear in any direction that is part of the intersection sectors. As a result, the shape and parameters of the intersection function ($FAI_{A \cap B}(i)$) can be any, but the value of this function cannot exceed the value of the minimum area of the figure, of those that form the scene.

For option 3 of the initial location of objects (Figure 8.1c), the values of $FAI_{AB}(i)$ will be equal to the value of part of a similar function for the case of close location of objects (Figure 8.1b).

According to ratio (8.3), the value of $Dist_{AB3}$ is negative (Figure 8.4). This fact must be taken into account when constructing the general function of the intersectional area ($FAI_{AB}(i)$). For the case of intersection of objects in formula (8.9), the value of $Dist_{AB}$ is equal to $Dist_{AB3}$.

8.2 Determining the Real Values of the Basic Parameters of the Scene Objects

Before we start processing the image, we can't know the shape of the scene objects, the area of each of them, or the distance between them. Only the actually obtained reference surface is available, which consists of the images of functions of the area of intersection ($FAI_{AB}(i)$) for shift directions from 0 to π. Since this set has the property of central symmetry, it is not necessary to

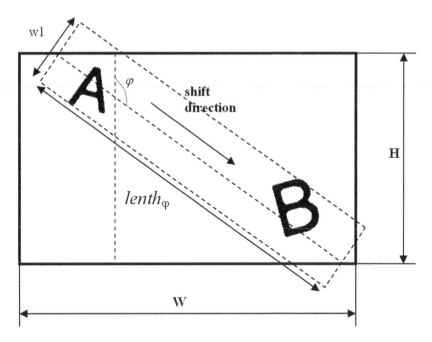

FIGURE 8.5
An example of the organization of the shift in a rectangular visual area.

determine the values of the functions for the shift directions from π to 2π. The total image area S_{0AB} is known. It is equal to the value of $FAI_{AB}(0)$ for any shift direction angle φ.

In order to determine the presence of a second object in the visual area in the first case of mutual location and distance $(Dist_{AB1})$ between objects, it is necessary to shift the copy of the image cyclically. Consider the case with a visual area of rectangular shape (Figure 8.5). In this case, one of the orthogonal directions (up) is selected for the zero shift direction.

The width of the visual area is equal W and the height is equal H. The length of the cyclic function of the area of intersection $(lenth_\varphi)$ for a certain direction φ will be equal to the following value.

$$lenth_\varphi = lenth_A + Dist_{AB} + lenth_B + Dist_{BA} \tag{8.11}$$

where

$lenth_A$ is the linear size of figure A for the selected shift direction φ;

$lenth_B$ is the linear size of figure B for the selected shift direction φ;

$Dist_{AB}$ is the distance from object A to object B;

$Dist_{BA}$ is the distance from object B to object A in the same shift direction.

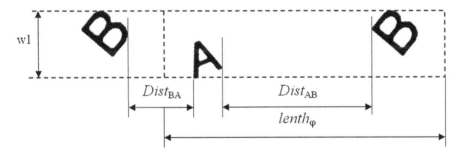

FIGURE 8.6
Calculation of the length of the offset band through the real linear sizes.

Scene analysis should solve the problem of finding the distance between objects ($Dist_{AB}$) in the selected direction.

The shift occurs in the offset band (Figure 8.6) of size w1 × $lenth_\varphi$, where the cyclic displacement of all elements of both objects by the length of the cycle is performed.

Also, the length of the offset band for the selected direction for the rectangular visual area can be found from the following system of equations.

$$lenth_\varphi = \begin{cases} H, when\,\varphi = 0\,or\,\varphi = \pi \\ \dfrac{H}{\cos\varphi}, when\left(tg\varphi < \dfrac{W}{H}\,and\,\varphi < \dfrac{\pi}{2}\right)or\left(ctg\varphi \left\langle \dfrac{W}{H}\,and\,\varphi \right\rangle \dfrac{\pi}{2}\right) \\ \dfrac{W}{\sin\varphi}, when\left(tg\varphi > \dfrac{W}{H}\,and\,\varphi < \dfrac{\pi}{2}\right)or\left(ctg\varphi > \dfrac{W}{H}\,and\,\varphi > \dfrac{\pi}{2}\right) \\ W, when\,\varphi = \dfrac{\pi}{2} \\ \sqrt{W^2 + H^2}, when\left(tg\varphi = \dfrac{W}{H}\,and\,\varphi < \dfrac{\pi}{2}\right)or\left(ctg\varphi = \dfrac{W}{H}\,and\,\varphi > \dfrac{\pi}{2}\right) \end{cases} \qquad (8.12)$$

Thus, with the help of the parameters of the visual area (W and H) and the shift direction φ we can determine the parameter $lenth_\varphi$.

Since the size, shape, spatial orientation of both objects of the scene, and the distance between them can be any (Figure 8.7), it will influence the shape of the general function of the area of intersection FAI_{AB}.

$$2\cdot\max\left(I_{max\,A}, I_{max\,B}\right) + Dist_{not\cap} \le lenth_\varphi \le 4\cdot\max\left(I_{max\,A}, I_{max\,B}\right) + Dist_{not\cap} \qquad (8.13)$$

where the value of the parameter $Dist_{not\cap}$ corresponds to the length of the areas of the shift of the image copies on which there is no intersection of

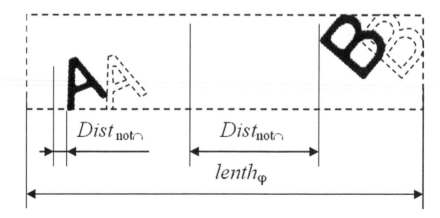

FIGURE 8.7
Calculation of the offset band length through the parameters of the function of the area of intersection.

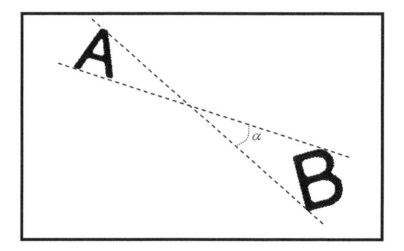

FIGURE 8.8
Sectors of intersection at cyclic shift.

the original figure (AB) with its copy. In the case of two objects that form a scene, the number of sections with no intersection can range from 0 to 3. The total width of the sections of intersection, depending on the shape, size, and relative position of the objects can vary from $2 \cdot \max (I_{maxA}, I_{maxB})$ to $4 \cdot \max (I_{maxA}, I_{maxB})$.

For certain directions of cyclic shift, there are sectors of intersection (Figure 8.8) which are similar to the variant of close location of objects (Figure 8.2).

The width of the offset band (w1) is equal to the height of the circumscribed rectangle for the shift direction φ (7.13). Since this direction is part of the intersection sector, the correction parameter $\Delta vert$ is greater than zero.

In directions which are not included in this intersection sector, the total intersection area function (FAI_{AB}) is formed by formula (8.8). It is necessary to create a set of the cyclic functions of the area of intersection for certain directions that are part of intersection sectors. For each graph of the $FAI_{AB}(i)$ function for a given shift direction, there must be several bursts on a section of $lenth_\varphi$. Their number is equal to the number of sections with no intersection (these sections alternate). A functions value that is greater than a certain limit value will be considered like the burst. This limit value is determined by the expected number of noise inclusions in the visual area.

Due to the variability of the parameters and location of the objects that form the scene, it is not possible to accurately determine the various ranges into which the general function of the intersection area is divided. However, in the case of cyclic shift for the time required for the shift to the distance $lenth_\varphi$, three events necessarily occur (Figure 8.9). The first is the intersection of figures A and B with their copies. The second is the intersection of figure A with a copy of figure B. The third is the intersection of figure B with a copy of figure A. Depending on the shape, size, and relative position of the two objects and the length of the offset band, the second and third events may change in order. It will be recalled that the first variant of the mutual arrangement of the figures that create the scene is considered (Figure 8.1a). The time ranges of these events may overlap. To determine some parameters of the

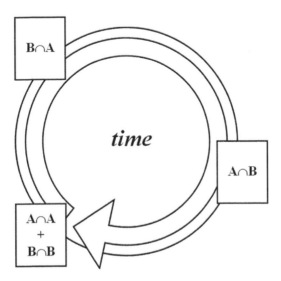

FIGURE 8.9
Three events when cyclically shifting a copy of image.

scene will allow the analysis of the total cyclic function of the cross-sectional area ($FAI_{AB}(i)$) for two objects in the selected direction.

Using the general cyclic function of the intersectional area, the following parameters can be determined.

1. The area of the initial figure (S_{0AB}). This value corresponds to the top of the maximum burst of the total FAI_{AB}.

$$S_{0AB} = \max\left(FAI_{AB}(i)\right) \qquad (8.14)$$

2. Maximum shift in the selected direction of the largest of the objects (I_{max}). This parameter is determined from the range corresponding to the time of the first event ($A \cap A + B \cap B$). It will be equal to the part of the burst with the maximum value of S_{0AB}. If the sizes of both objects are smaller than the distances between them ($Dist_{AB}$ and $Dist_{BA}$), then this part is equal to half of such a burst (the first variant of the mutual intersection). If the length of the offset band is small, then the range of the second or third event may intersect with the time range of the first event. This is the second variant of the intersection, which arose due to the cyclic shift. Then the maximum shift will be equal to the minimum distance between the coordinate with zero value $FAI_{AB}(i)$ (x_{zero} – number of such points on the graph of the function can be from 0 to $lenth_{\varphi}$) and the coordinate x_{max} with the maximum value of this function (S_{0AB}). The coordinate with the maximum value can be one, or if both objects are the same and equally spatially oriented, then such points on the graph of the function can be two or three (intersection of regularly located identical objects).

$$I_{max} = \max\left(I_{maxA}, I_{maxB}\right) = \min\left(\left|x_{max} - x_{zero}\right|\right) \qquad (8.15)$$

3. The maximum value of the function of the mutual intersection of the scene objects for a given direction of shift ($FAI_{A \cap B}(i)$). This value corresponds to the maximum value of the two smaller bursts, which are equal to each other and reflect the values for the intervals of the second and third event ($A \cap B$ or $B \cap A$).

$$FAI_{A \cap B}(i) \le \min\left(S_{0A}, S_{0B}\right) \qquad (8.16)$$

4. The length of the intersection range of two objects is equal to the width of one of the two smaller bursts, which correspond to the intervals of the second and third event ($lenth_{A \cap B}$).

$$0 \le lenth_{A \cap B} \le 2 \cdot \max\left(I_{maxA}, I_{maxB}\right) \qquad (8.17)$$

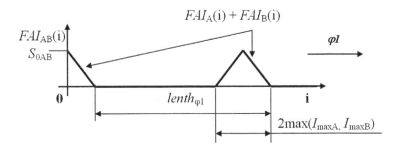

FIGURE 8.10
Schematic representation of the cyclic function of the area of intersection for direction outside the intersection sectors.

Graphs of cyclic intersection functions for a range of directions that are not included in the intersection sectors have only one burst of values (Figure 8.10). This graph shows the first event (intersection of scene objects with their copies).

From the data in this graph, formula (8.14) determines the total area of the scene objects (S_{0AB}). In addition, if $\varphi1$ is chosen as the direction of the shift, where $\phi1 = \phi \pm \dfrac{\pi}{2}$, then according to formula (7.13), it is possible to calculate the parameter of the height of the circumscribed rectangle for the object AB for the direction $\varphi1$. This height is created by the linear sizes of both parts of the scene in the direction φ and the distance between them. In this case, the distance between the objects $Dist_{AB}$ for the direction φ is equal to the modulus of the parameter $\Delta vert$. The real parameters of the circumscribed rectangles can be obtained by calculating the distance from the scene objects to the edges of the visual area. For directions that are not included in the intersection sectors, the following ratio is valid.

$$Dist_{AB} = \left|\Delta vert_{\varphi1}\right| = \left|B_{0\phi1} - S_{0AB} + FAI_{\varphi1}(1)\right| \tag{8.18}$$

where $B_{0\varphi1}$ is the height of the rectangle bounding the scene when shifted in the direction $\varphi1$.

If there is only one symmetrical burst in the range of the general intersection function in the selected direction φ, then in this case the first variant of the mutual location of the objects of the scene takes place, or in this direction outside the intersection sector of the images.

The distance ($Dist_{AB}$) from figure A to figure B found by formula (8.18) is the solution of the first subtask of scene analysis. On a three-dimensional map of the surrounding world, an intelligent system can capture the relative positions of two objects. The shape and size of both objects remain unknown.

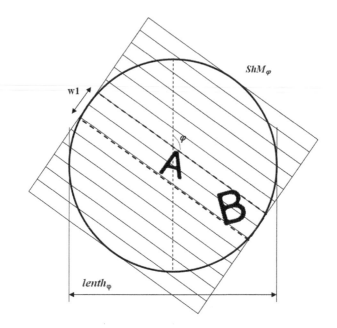

FIGURE 8.11

An example of the organization of the shift in the general circular visual area.

Offset bands are parts of shift matrices ShM_φ (Figure 8.11), which are a set of cyclic shift registers. The sizes of the sides of the shift matrices (h_{SM}) must provide image processing of any size that does not exceed the parameters of the visual area. For a rectangular visual area, they are as follows.

$$h_{SM} = \max\left(lenth_\varphi\right) = \max\left(w1\right) \tag{8.19}$$

To simplify the calculations, the dimensions of the shift matrix should be the same. That is, each such matrix is a square. Inside each square shift matrix there is a certain circular area that is tangent to the sides of the matrix (Figure 8.11). These regions are the common visual area for the base matrix and all available shift matrices. Images that are analyzed should be smaller than this area.

This form is inherent in the organs of vision of living beings. Usually, biological beings focus on one of the objects in the scene. It is located in the center of the visual area. The parameters of all other components of the scene are calculated relative to it. The central element of the scene can be replaced by another. This will recalculate the distance parameters between all objects. After obtaining the location data of object A relative to object B (when B is the central element), we can check the calculated parameters of the location of object B relative to object A (when A is the central element).

Such an organization of the visual area is convenient because the size of the *lenth*$_\varphi$ is always equal to the diameter of the circle. The parameters of the shift matrices (*ShM*$_\varphi$) must be the same size. That is, the parameters of cyclic functions for all directions will be constant. This fact can greatly simplify the technical implementation of the system of forming a set of cyclic functions of the area of intersection of the elements of the scene for each possible direction of shift of the image copy.

8.3 Determining the Shape of the Objects of Scene

To determine the shape of the objects of research, it is necessary to compare the obtained functions of the area of intersection $FAI_{AB}(i)$ in the distance from 0 to max(I_{maxA}, I_{maxB}) with combinations of pairs of reference *FAIs* (formula 38). These studies should be performed for shift directions that are not included in the intersection sectors. For a scene of two objects, these directions can be determined by the shape of the cyclic *FAIs*. These directions are determined by the presence of only one burst of the general cyclic function on each section of length *lenth*$_\varphi$ (Figure 8.10).

Let the number of etalons in the database be equal to m. Let the number of directions for which the *FAIs* of each etalon is created be equal to n. The possible real linear size for each object for each direction φ can take values from 0 to max($I_{maxA\varphi}$, $I_{maxB\varphi}$). The maximum value of this parameter is equal to *lenth*$_\varphi$.

Then the number of possible pairs (*Npp*) of the functions of the area of intersection of the objects to be summed may be as follows.

$$Npp = \left(m \cdot n \cdot lenth_\varphi \right)^2 \tag{8.20}$$

In addition, in each case it is necessary to scale the reference functions of the area of intersection and compare the functions. These processes require significant computing power.

To reduce the number of pairs of functions for summation, as in image recognition (Bilan & Yuzhakov, 2018), we will divide the process into two stages (fast and detail). On a fast stage, a variety of additional parameters can be used for the search, which are stored in the reference database for each etalon for each shift direction. The purpose of this process is to reduce the number of possible pairs of functions of the area of intersection for further comparison.

For example, you can use a parameter such as the integral coefficient.

$$k_\varphi = \frac{\int_0^{I_{max}} FAI(i)\,di}{S_0 \cdot I_{max}} = const \tag{8.21}$$

where φ is the direction for which the *FAI* is determined, i and $I_{\max}$ are the shift and the maximum shift in this direction, respectively. In this formula, S_0 acts as a one-dimensional quantity. This is the height of the location of the *FAI*(i) graph. Then the parameter k_φ is dimensionless.

Using the parameters of the obtained cyclic functions of the area of intersection of the scene, we can calculate their corresponding integral coefficients. They are equal to the area of the half of the burst, divided by the half of the multiplication of the maximum value and the length of the burst, for each direction of shift, which is not included in the intersection sectors.

$$k_{AB\phi} = \frac{\displaystyle\int_0^{\max\left(I_{\max A\phi}, I_{\max B\phi}\right)} FAI_{AB\phi}(i)\,di}{S_{0AB} \cdot \max\left(I_{\max A\phi}, I_{\max B\phi}\right)}$$

$$= \frac{\displaystyle\int_0^{\max\left(I_{\max A\phi}, I_{\max B\phi}\right)} \left(SC_{A\psi 1}{}^2 \cdot FAI_{A\psi 1}(i) + SC_{B\psi 1}{}^2 \cdot FAI_{B\psi 1}(i)\right)\,di}{S_{0AB} \cdot \max\left(I_{\max A\phi}, I_{\max B\phi}\right)} \tag{8.22}$$

In this case, the angle φ is the direction of the image copy shift selected for analysis. The part of the function $FAI_{AB\varphi}(i)$ corresponds to the values of the unit burst range of the total cyclic function of the area of intersection for this direction. Angles $A\psi 1$ and $B\psi 1$ are the directions for constructing the functions of the area of intersection for objects A and B from the reference database that are currently selected for verification. We do not know the vertical orientation of the objects, so these angles can have any meaning. The intersection functions of the reference objects should be scaled using the scaling factors ($SC_{A\psi 1}$ and $SC_{B\psi 1}$). The calculation of these coefficients is described subsequently.

The division of the function $FAI_{AB\varphi}(i)$ into two parts occurs due to the application of formula 38.

Consider a certain other direction $\varphi 1$ for the formation of a cyclic *FAI* of scene. It should also not be part of the intersection sectors. We do not know the values of the angles for the directions $A\psi 2$ and $B\psi 2$. However, we know exactly the following.

$$\phi 1 - \phi = A\psi 2 - A\psi 1 = B\psi 2 - B\psi 1 \tag{8.23}$$

Using the values of the reference maximum shifts for the directions $A\psi 1$ and $B\psi 1$ and $A\psi 2$ and $B\psi 2$, it is possible to determine certain proportionality coefficients (PC_A and PC_B), which will be used in the fast stage of comparison. These coefficients reflect the relationship between certain functions of the area of intersection of figures with the same shape. These parameters do not depend on the real size of the objects.

$$PC_A = \frac{I_{\max A\psi 2}}{I_{\max A\psi 1}} = \frac{I_{\max A\phi 1}}{I_{\max A\phi}} \qquad (8.24)$$

$$PC_B = \frac{I_{\max B\psi 2}}{I_{\max B\psi 1}} = \frac{I_{\max B\phi 1}}{I_{\max B\phi}} \qquad (8.25)$$

The real *FAI* parameters of an object may differ from the reference ones with the same form. There is a need to apply certain scaling factors ($SC_{A\psi 1}$ and $SC_{B\psi 1}$). These values will determine the possible real values of the initial areas (S_{0A} and S_{0B}). We will calculate the scaling factors for the reference directions $A\psi 1$ and $B\psi 1$. For directions $A\psi 2$ and $B\psi 2$ they should be similar.

$$SC_{A\psi 1} = \frac{I_{\max A\phi}}{I_{\max A\psi 1}} = SC_{A\psi 2} = \frac{I_{\max A\phi 1}}{I_{\max A\psi 2}} \qquad (8.26)$$

$$SC_{B\psi 1} = \frac{I_{\max B\phi}}{I_{\max B\psi 1}} = SC_{B\psi 2} = \frac{I_{\max B\phi 1}}{I_{\max B\psi 2}} \qquad (8.27)$$

Using the scaling factors, we calculate the possible real values of the initial areas of the study objects (S_{0A} and S_{0B}). They are equal to the multiplication of the squares of the scaling factors and the values of the reference initial areas of the selected objects ($S_{0A\psi 1}$ and $S_{0B\psi 1}$). These values of scaling are the same for all directions ψ of each selected etalon.

$$S_{0A} = S_{0A\psi 1} \cdot SC_{A\psi 1}^{2} = S_{0A\psi 2} \cdot SC_{A\psi 1}^{2} \qquad (8.28)$$

$$S_{0B} = S_{0B\psi 1} \cdot SC_{B\psi 1}^{2} = S_{0B\psi 2} \cdot SC_{B\psi 1}^{2} \qquad (8.29)$$

The parameter of the area in this case is two-dimensional, so in these formulas take the squares of the scaling factors.

If we take into account formula (8.21) for the reference directions and the graph of the total cyclic function of the area of intersection for the direction φ, which is not part of the intersection sector, then for object A we can confirm the following.

$$SC_{A\psi 1}^{2} \cdot \int_{0}^{\max(I_{\max A\phi}, I_{\max B\phi})} FAI_{A\psi 1}(i)\,di = \int_{0}^{I_{\max A\phi}} FAI_{A\phi}(i)\,di = k_{A\psi 1} \cdot S_{0A} \cdot I_{\max A\phi} \qquad (8.30)$$

It should be borne in mind that $k_{A\phi} = k_{A\psi 1} = const$, and the value of the function $FAI_{A\psi 1}(i) = 0$ in the range of arguments from $I_{\max A\varphi}$ to $\max(I_{\max A\varphi}, I_{\max B\varphi})$, if the linear size of the object A (*lenth$_A$*) is less than the similar size of the object B (*lenth$_B$*) for this direction of shift. In this formula, S_{0A} again acts as a one-dimensional quantity as part of formula (8.21) for calculating the integral coefficient.

A similar statement is possible with respect to the functions of object B.

$$SC_{B\psi 1}{}^2 \cdot \int_0^{\max\left(I_{\max A\phi}, I_{\max B\phi}\right)} FAI_{B\psi 1}(i)\,di = \int_0^{I_{\max B\phi}} FAI_{B\phi}(i)\,di = k_{B\psi 1} \cdot S_{0B} \cdot I_{\max B\phi} \quad (8.31)$$

Due to the above formula (8.22) can be transformed as follows.

$$k_{AB\phi} = \frac{k_{A\psi 1} \cdot S_{0A} \cdot I_{\max A\phi} + k_{B\psi 1} \cdot S_{0B} \cdot I_{\max B\phi}}{\left(S_{0A} + S_{0B}\right) \cdot \max\left(I_{\max A\phi}, I_{\max B\phi}\right)} \quad (8.32)$$

This will express the value of the initial area of one scene object through the initial area of another scene object.

$$\frac{S_{0A}}{k_{B\psi 1} \cdot I_{\max B\phi} - k_{AB\phi} \cdot \max\left(I_{\max A\phi}, I_{\max B\phi}\right)} = \frac{S_{0B}}{k_{AB\phi} \cdot \max\left(I_{\max A\phi}, I_{\max B\phi}\right) - k_{A\psi 1} \cdot I_{\max A\phi}} \quad (8.33)$$

We can also express the values of the initial areas of the scene objects through the initial area of the entire image.

$$S_{0A} = S_{0AB} \cdot \frac{k_{AB\phi} \cdot \max\left(I_{\max A\phi}, I_{\max B\phi}\right) - k_{B\psi 1} \cdot I_{\max B\phi}}{k_{A\psi 1} \cdot I_{\max A\phi} - k_{B\psi 1} \cdot I_{\max B\phi}} \quad (8.34)$$

The formula for determining $S_{0B\phi}$ is formed similarly.

$$S_{0B} = S_{0AB} \cdot \frac{k_{AB\phi} \cdot \max\left(I_{\max A\phi}, I_{\max B\phi}\right) - k_{A\psi 1} \cdot I_{\max A\phi}}{k_{B\psi 1} \cdot I_{\max B\phi} - k_{A\psi 1} \cdot I_{\max A\phi}} \quad (8.35)$$

The values of the parameters S_{0AB}, $\max(I_{\max A\phi}, I_{\max B\phi})$, and $k_{AB\phi}$ can be obtained by analyzing the general cyclic function of the area of intersection of the scene for the shift direction ϕ. We do not know the value of other parameters. These are in some way scalable values of reference formulas. For example, $I_{\max A\phi}$ is a scalable $I_{\max A\psi 1}$, and $S_{0A\phi}$ is a scalable $S_{0A\psi 1}$. The values of the integral coefficients do not change when scaling the functions.

Consider a certain other direction $\phi 1$ for the formation of a cyclic *FAI* of scene. It must also be outside the intersection sectors. For this direction of the cyclic shift ($\phi 1$), we can write formulas similar to formulas (8.34) and (8.35).

$$S_{0A} = S_{0AB} \cdot \frac{k_{AB\phi1} \cdot \max\left(I_{\max A\phi1}, I_{\max B\phi1}\right) - k_{B\psi2} \cdot I_{\max B\phi1}}{k_{A\psi2} \cdot I_{\max A\phi1} - k_{B\psi2} \cdot I_{\max B\phi1}} \tag{8.36}$$

$$S_{0B} = S_{0AB} \cdot \frac{k_{AB\phi1} \cdot \max\left(I_{\max A\phi1}, I_{\max B\phi1}\right) - k_{A\psi2} \cdot I_{\max A\phi1}}{k_{B\psi2} \cdot I_{\max B\phi1} - k_{A\psi2} \cdot I_{\max A\phi1}} \tag{8.37}$$

We also know that the values of the real initial areas for the whole scene (S_{0AB}) and its individual elements (S_{0A} and S_{0B}) constant for all directions of formation of cyclic functions of the area of intersection of the scene. Then we can combine formulas (8.34) and (8.35), (8.36) and (8.37) in pairs.

$$\begin{aligned}
&\frac{k_{AB\phi} \cdot \max\left(I_{\max A\phi}, I_{\max B\phi}\right) - k_{B\psi1} \cdot I_{\max B\phi}}{k_{A\psi1} \cdot I_{\max A\phi} - k_{B\psi1} \cdot I_{\max B\phi}} \\
&= \frac{k_{AB\phi1} \cdot \max\left(I_{\max A\phi1}, I_{\max B\phi1}\right) - k_{B\psi2} \cdot I_{\max B\phi1}}{k_{A\psi2} \cdot I_{\max A\phi1} - k_{B\psi2} \cdot I_{\max B\phi1}}
\end{aligned} \tag{8.38}$$

$$\begin{aligned}
&\frac{k_{AB\phi} \cdot \max\left(I_{\max A\phi}, I_{\max B\phi}\right) - k_{A\psi1} \cdot I_{\max A\phi}}{k_{B\psi1} \cdot I_{\max B\phi} - k_{A\psi1} \cdot I_{\max A\phi}} \\
&= \frac{k_{AB\phi1} \cdot \max\left(I_{\max A\phi1}, I_{\max B\phi1}\right) - k_{A\psi2} \cdot I_{\max A\phi1}}{k_{B\psi2} \cdot I_{\max B\phi1} - k_{A\psi2} \cdot I_{\max A\phi1}}
\end{aligned} \tag{8.39}$$

We do not know the real values of the maximum shifts for individual objects ($I_{maxA\varphi}$, $I_{maxB\varphi}$, $I_{maxA\varphi1}$, $I_{maxB\varphi1}$). There is a need to search for a number of options.

There are three options for the linear sizes of objects for each direction of the cyclic function of the area of intersection.

1. $\max(I_{maxA}, I_{maxB}) = I_{maxA} = I_{maxB}$ (objects are equal in size in the direction of shift);

2. $\max(I_{maxA}, I_{maxB}) = I_{maxA} > I_{maxB}$ (object A is larger than object B in size in the direction of shift);

3. $\max(I_{maxA}, I_{maxB}) = I_{maxB} > I_{maxA}$ (object B is larger than object A in size in the direction of shift).

For two directions of formation of cyclic FAIs (φ and $\varphi1$), there are nine variants of transformation of formulas (8.38) and (8.39) (from 1.1 to 3.3). For example, the numbering of option 2.3 means that the equation $\max(I_{maxA}, I_{maxB}) = I_{maxA} > I_{maxB}$ is true for the shift direction φ and the equation $\max(I_{maxA}, I_{maxB}) = I_{maxB} > I_{maxA}$ is true for the shift direction $\varphi1$. That is, the first digit of the variant of formation of cyclic functions of the intersection area is equal to the number of the variant of linear sizes of objects for the direction φ. The second digit of the variant of formation of cyclic functions of the area of intersection is equal to the variant number of the linear sizes of objects for the direction $\varphi 1$.

The simplest option is option 1.1. That is, for both directions, the values of the maximum shifts are the same and equal to the maximum shifts of both objects.

$$\frac{k_{AB\phi} - k_{B\psi 1}}{k_{A\psi 1} - k_{B\psi 1}} = \frac{k_{AB\phi 1} - k_{B\psi 2}}{k_{A\psi 2} - k_{B\psi 2}} \tag{8.40}$$

$$\frac{k_{AB\phi} - k_{A\psi 1}}{k_{B\psi 1} - k_{A\psi 1}} = \frac{k_{AB\phi 1} - k_{A\psi 2}}{k_{B\psi 2} - k_{A\psi 2}} \tag{8.41}$$

Recall that the values of $k_{AB\varphi}$ and $k_{AB\varphi 1}$ can be calculated by analyzing the cyclic functions of the area of intersection of the scene for these directions. Other parameters are obtained from the reference database. The possible real values of the maximum shifts of objects in the corresponding directions are obtained from cyclic *FAIs*. They are equal to $\max(I_{\max A\varphi}, I_{\max B\varphi})$ and $\max(I_{\max A\varphi 1}, I_{\max B\varphi 1})$. This is half the range of unit bursts of the graphs of the functions $FAI_{AB}(i)$ for the directions φ and $\varphi 1$. Substituting the values of the integral coefficients of the standards for the directions $\psi 1$ and $\psi 2$ into the obtained functions, it is possible to draw a conclusion about the possibility of the identity of the shapes of the reference and real objects. If equations 8.40 and 8.41 are truth, then we can proceed to a detailed stage of comparison (comparison of the functions of the area of intersection of the reference and real objects). To do this, the *FAIs* of both standards are scaled using the scaling factors (formulas (8.26) and (8.27)), summed (formula (8.8)), compared with part (half) of the non-zero sections of the real general cyclic functions of the area of intersection for the directions φ and $\varphi 1$. If a match occurs, the scene objects are assumed to be identical to the selected reference objects. The real parameters of the scene objects are obtained by scaling the parameters of the selected standards.

For variants of linear sizes of objects 1.2 and 1.3, formula 8.38 is transformed as follows. Hereinafter, the use of formula 8.39 is not appropriate. It gives a result similar to formula 8.38.

For option 1.2:

$$\frac{k_{AB\phi} - k_{B\psi 1}}{k_{A\psi 1} - k_{B\psi 1}} = \frac{k_{AB\phi 1} \cdot \max\left(I_{\max A\phi 1}, I_{\max B\phi 1}\right) - k_{B\psi 2} \cdot I_{\max B\phi 1}}{k_{A\psi 2} \cdot \max\left(I_{\max A\phi 1}, I_{\max B\phi 1}\right) - k_{B\psi 2} \cdot I_{\max B\phi 1}} \tag{8.42}$$

From this formula, we obtain the real value of the maximum shift of object B for the direction $\varphi 1$.

$$I_{\max B\phi 1} = \frac{\left(\left(k_{A\psi 1} - k_{B\psi 1}\right) \cdot k_{AB\phi 1} - \left(k_{AB\phi} - k_{B\psi 1}\right) \cdot k_{A\psi 2}\right) \cdot \max\left(I_{\max A\phi 1}, I_{\max B\phi 1}\right)}{\left(k_{AB\phi} - k_{A\psi 1}\right) \cdot k_{B\psi 2}} \tag{8.43}$$

The real values of the parameters $I_{maxA\varphi}$, $I_{maxB\varphi}$, $I_{maxA\varphi1}$ are obtained from the general cyclic functions of the area of intersection for the shift directions φ and $\varphi1$ in accordance with the variant number of the linear sizes of objects (1.2).

For option 1.3:

$$\frac{k_{AB\phi} - k_{B\psi1}}{k_{A\psi1} - k_{B\psi1}} = \frac{k_{AB\phi1} \cdot \max\left(I_{\max A\phi1}, I_{\max B\phi1}\right) - k_{B\psi2} \cdot \max\left(I_{\max A\phi1}, I_{\max B\phi1}\right)}{k_{A\psi2} \cdot I_{\max A\phi1} - k_{B\psi2} \cdot \max\left(I_{\max A\phi1}, I_{\max B\phi1}\right)} \tag{8.44}$$

From this formula, we obtain the real value of the maximum shift of object A for the direction $\varphi1$.

$$I_{\max A\phi1} = \frac{\left(\left(k_{A\psi1} - k_{B\psi1}\right) \cdot k_{AB\phi1} + \left(k_{AB\phi} - k_{A\psi1}\right) \cdot k_{B\psi2}\right) \cdot \max\left(I_{\max A\phi1}, I_{\max B\phi1}\right)}{\left(k_{AB\phi} - k_{B\psi1}\right) \cdot k_{A\psi2}} \tag{8.45}$$

The real values of the parameters $I_{maxA\varphi}$, $I_{maxB\varphi}$, $I_{maxB\varphi1}$ are obtained from the general cyclic functions of the area of intersection for the shift directions φ and $\varphi1$ in accordance with the variant number of the linear sizes of objects (1.3).

Options 2.1 and 3.1 are calculated similarly to options 1.2 and 1.3. Only calculations are performed on the basis of cyclic *FAIs* for the direction φ. Then for option 2.1:

$$\frac{k_{AB\phi} \cdot \max\left(I_{\max A\phi}, I_{\max B\phi}\right) - k_{B\psi1} \cdot I_{\max B\phi}}{k_{A\psi1} \cdot \max\left(I_{\max A\phi}, I_{\max B\phi}\right) - k_{B\psi1} \cdot I_{\max B\phi}} = \frac{k_{AB\phi1} - k_{B\psi2}}{k_{A\psi2} - k_{B\psi2}} \tag{8.46}$$

$$I_{\max B\phi} = \frac{\left(\left(k_{A\psi2} - k_{B\psi2}\right) \cdot k_{AB\phi} - \left(k_{AB\phi1} - k_{B\psi2}\right) \cdot k_{A\psi1}\right) \cdot \max\left(I_{\max A\phi}, I_{\max B\phi}\right)}{\left(k_{AB\phi1} - k_{A\psi2}\right) \cdot k_{B\psi1}} \tag{8.47}$$

For option 3.1:

$$\frac{k_{AB\phi} \cdot \max\left(I_{\max A\phi}, I_{\max B\phi}\right) - k_{B\psi1} \cdot \max\left(I_{\max A\phi}, I_{\max B\phi}\right)}{k_{A\psi1} \cdot I_{\max A\phi} - k_{B\psi1} \cdot \max\left(I_{\max A\phi}, I_{\max B\phi}\right)} = \frac{k_{AB\phi1} - k_{B\psi2}}{k_{A\psi2} \cdot - k_{B\psi2}} \tag{8.48}$$

$$I_{\max A\phi} = \frac{\left(\left(k_{A\psi2} - k_{B\psi2}\right) \cdot k_{AB\phi} + \left(k_{AB\phi1} - k_{A\psi2}\right) \cdot k_{B\psi1}\right) \cdot \max\left(I_{\max A\phi}, I_{\max B\phi}\right)}{\left(k_{AB\phi1} - k_{B\psi2}\right) \cdot k_{A\psi1}} \tag{8.49}$$

When the values of the parameters $I_{maxA\varphi}$, $I_{maxA\varphi1}$, $I_{maxB\varphi}$ and $I_{maxB\varphi1}$ are known, we check the proportionality coefficients (formulas (8.24) and (8.25)). If both equations are not true, then we move on to consider other directions of the etalons, or to consider other etalons. If one of these equations is not true, then we proceed to consider another direction of the etalon to which it relates, or to consider another etalon. If equations (8.24) and (8.25) are true, then we proceed to the detailed stage of comparing the sum of the scaled functions of the area of intersection of the reference objects with the non-zero parts of the general cyclic functions of the area of intersection for the directions φ and $\varphi 1$.

The initial formulas for options 2.2 and 3.3 include two unknowns.
For option 2.2:

$$\frac{k_{AB\phi} \cdot \max\left(I_{\max A\phi}, I_{\max B\phi}\right) - k_{B\psi 1} \cdot I_{\max B\phi}}{k_{A\psi 1} \cdot \max\left(I_{\max A\phi}, I_{\max B\phi}\right) - k_{B\psi 1} \cdot I_{\max B\phi}} = \frac{k_{AB\phi 1} \cdot \max\left(I_{\max A\phi 1}, I_{\max B\phi 1}\right) - k_{B\psi 2} \cdot I_{\max B\phi 1}}{k_{A\psi 2} \cdot \max\left(I_{\max A\phi 1}, I_{\max B\phi 1}\right) - k_{B\psi 2} \cdot I_{\max B\phi 1}}$$

$$(8.50)$$

In this case, the values of $I_{maxB\varphi}$ and $I_{maxB\varphi 1}$ are unknown.
For option 3.3:

$$\frac{k_{AB\phi} \cdot \max\left(I_{\max A\phi}, I_{\max B\phi}\right) - k_{B\psi 1} \cdot \max\left(I_{\max A\phi}, I_{\max B\phi}\right)}{k_{A\psi 1} \cdot I_{\max A\phi} - k_{B\psi 1} \cdot \max\left(I_{\max A\phi}, I_{\max B\phi}\right)}$$

$$= \frac{k_{AB\phi 1} \cdot \max\left(I_{\max A\phi 1}, I_{\max B\phi 1}\right) - k_{B\psi 2} \cdot \max\left(I_{\max A\phi 1}, I_{\max B\phi 1}\right)}{k_{A\psi 2} \cdot I_{\max A\phi 1} - k_{B\psi 2} \cdot \max\left(I_{\max A\phi 1}, I_{\max B\phi 1}\right)} \quad (8.51)$$

In this case, the values of $I_{maxA\varphi}$ and $I_{maxA\varphi 1}$ are unknown.

For these two cases, the fact that the values of the real sizes should be proportional to the reference should be used. That is, $I_{maxA\varphi}$ is proportional to $I_{maxA\psi 1}$, $I_{maxA\varphi 1}$ is proportional to $I_{maxA\psi 2}$, $I_{maxB\varphi}$ is proportional to $I_{maxB\psi 1}$, $I_{maxB\varphi 1}$ is proportional to $I_{maxB\psi 2}$. There is a need to apply the proportionality coefficient (PC_A and PC_B).

For better visibility of the display of formulas, we made the following changes.

$$\max\left(I_{\max A\phi}, I_{\max B\phi}\right) = M\phi \tag{8.52}$$

$$\max\left(I_{\max A\phi 1}, I_{\max B\phi 1}\right) = M\phi 1 \tag{8.53}$$

Then formulas 8.50 and 8.51 are converted to the following.
For option 2.2:

$$\frac{k_{AB\phi} \cdot M\phi - k_{B\psi 1} \cdot I_{\max B\phi}}{k_{A\psi 1} \cdot M\phi - k_{B\psi 1} \cdot I_{\max B\phi}} = \frac{k_{AB\phi 1} \cdot M\phi 1 - k_{B\psi 2} \cdot I_{\max B\phi} \cdot PC_B}{k_{A\psi 2} \cdot M\phi 1 - k_{B\psi 2} \cdot I_{\max B\phi} \cdot PC_B} \tag{8.54}$$

$$I_{\max B\phi} = \frac{\left(k_{AB\phi 1} \cdot k_{A\psi 1} - k_{AB\phi} \cdot k_{A\psi 2}\right) \cdot M\phi \cdot M\phi 1}{\left(k_{AB\phi 1} - k_{A\psi 2}\right) \cdot k_{B\psi 1} \cdot M\phi 1 + \left(k_{A\psi 1} - k_{AB\phi}\right) \cdot k_{B\psi 2} \cdot M\phi \cdot PC_B} \tag{8.55}$$

For option 3.3:

$$\frac{k_{AB\phi} \cdot M\phi - k_{B\psi 1} \cdot M\phi}{k_{A\psi 1} \cdot I_{\max A\phi} - k_{B\psi 1} \cdot M\phi} = \frac{k_{AB\phi 1} \cdot M\phi 1 - k_{B\psi 2} \cdot M\phi 1}{k_{A\psi 2} \cdot I_{\max A\phi} \cdot PC_A - k_{B\psi 2} \cdot M\phi 1} \tag{8.56}$$

$$I_{\max A\phi} = \frac{\left(k_{AB\phi} \cdot k_{B\psi 2} - k_{AB\phi 1} \cdot k_{B\psi 1}\right) \cdot M\phi \cdot M\phi 1}{\left(k_{B\psi 2} - k_{AB\phi 1}\right) \cdot k_{A\psi 1} \cdot M\phi 1 + \left(k_{AB\phi} - k_{B\psi 1}\right) \cdot k_{A\psi 2} \cdot M\phi \cdot PC_A} \tag{8.57}$$

We use in formulas (8.54)÷(8.57) proportionality coefficients (PC_A and PC_B), which are calculated on the basis of the parameters of the selected standards for the selected directions ($A\psi 1$ and $B\psi 1$, $A\psi 2$ and $B\psi 2$). When the values of the parameters $I_{\max A\phi}$, $I_{\max A\phi 1}$, $I_{\max B\phi}$ and $I_{\max B\phi 1}$ are known, we check the proportionality coefficients, calculating these coefficients based on these parameters. If equations (8.24) and (8.25) are truth, then we proceed to a detailed comparison of scaled *FAIs*.

All previous calculations were performed on the basis of formulas 8.34 and 8.36. Similar calculations can be performed on the basis of formulas (8.35) and (8.37).

For option 2.3, the search for the values of possible real maximum shifts of objects is performed using the following system of equations.

$$\begin{cases} \dfrac{k_{AB\phi} \cdot M\phi - k_{B\psi 1} \cdot I_{\max B\phi}}{k_{A\psi 1} \cdot M\phi - k_{B\psi 1} \cdot I_{\max B\phi}} = \dfrac{k_{AB\phi 1} \cdot M\phi 1 - k_{B\psi 2} \cdot M\phi 1}{k_{A\psi 2} \cdot I_{\max A\phi 1} - k_{B\psi 2} \cdot M\phi 1} \\[4mm] \dfrac{k_{AB\phi} \cdot M\phi - k_{A\psi 1} \cdot M\phi}{k_{B\psi 1} \cdot I_{\max B\phi} - k_{A\psi 1} \cdot M\phi} = \dfrac{k_{AB\phi 1} \cdot M\phi 1 - k_{A\psi 2} \cdot I_{\max A\phi 1}}{k_{B\psi 2} \cdot M\phi 1 - k_{A\psi 2} \cdot I_{\max A\phi 1}} \end{cases} \tag{8.58}$$

This system can also be transformed to express the values of possible real maximum shifts of objects through each other.

$$\begin{cases} I_{\max B\phi} = \dfrac{M\phi \cdot \left(\left(\left(k_{AB\phi} - k_{A\psi 1}\right) \cdot k_{B\psi 2} + k_{AB\phi 1} \cdot k_{A\psi 1}\right) \cdot M\phi 1 - k_{AB\phi} \cdot k_{A\psi 2} \cdot I_{\max A\phi 1}\right)}{\left(k_{AB\phi 1} \cdot M\phi 1 - k_{A\psi 2} \cdot I_{\max A\phi 1}\right) \cdot k_{B\psi 1}} \\[4mm] I_{\max A\phi 1} = \dfrac{M\phi 1 \cdot \left(\left(\left(k_{AB\phi 1} - k_{B\psi 2}\right) \cdot k_{A\psi 1} + k_{AB\phi} \cdot k_{B\psi 2}\right) \cdot M\phi - k_{AB\phi 1} \cdot k_{B\psi 1} \cdot I_{\max B\phi}\right)}{\left(k_{AB\phi} \cdot M\phi - k_{B\psi 1} \cdot I_{\max B\phi}\right) \cdot k_{A\psi 2}} \end{cases} \tag{8.59}$$

In addition, to find the value of possible maximum shifts, we use the proportionality coefficients (PC_A and PC_B) of formulas (8.24) and (8.25).

$$I_{\max A\phi 1} = PC_A \cdot I_{\max A\phi} = PC_A \cdot M\phi \tag{8.60}$$

$$I_{\max B\phi} = \frac{I_{\max B\phi 1}}{PC_B} = \frac{M\phi 1}{PC_B} \tag{8.61}$$

For option 3.2, the search for the values of possible real maximum shifts of objects is carried out similarly to option 2.3.

$$\begin{cases} \dfrac{k_{AB\phi} \cdot M\phi - k_{B\psi1} \cdot M\phi}{k_{A\psi1} \cdot I_{\max A\phi} - k_{B\psi1} \cdot M\phi} = \dfrac{k_{AB\phi1} \cdot M\phi1 - k_{B\psi2} \cdot I_{\max B\phi1}}{k_{A\psi2} \cdot M\phi1 - k_{B\psi2} \cdot I_{\max B\phi1}} \\[4mm] \dfrac{k_{AB\phi} \cdot M\phi - k_{A\psi1} \cdot I_{\max A\phi}}{k_{B\psi1} \cdot M\phi - k_{A\psi1} \cdot I_{\max A\phi}} = \dfrac{k_{AB\phi1} \cdot M\phi1 - k_{A\psi2} \cdot M\phi1}{k_{B\psi2} \cdot I_{\max B\phi1} - k_{A\psi2} \cdot M\phi1} \end{cases} \tag{8.62}$$

This system can also be transformed to express the values of possible real maximum shifts of objects through each other.

$$\begin{cases} I_{\max A\phi} = \dfrac{M\phi \cdot \left(\left(\left(k_{AB\phi} - k_{B\psi1}\right) \cdot k_{A\psi2} + k_{AB\phi1} \cdot k_{B\psi1}\right) \cdot M\phi1 - k_{AB\phi} \cdot k_{B\psi2} \cdot I_{\max B\phi1}\right)}{\left(k_{AB\phi1} \cdot M\phi1 - k_{B\psi2} \cdot I_{\max B\phi1}\right) \cdot k_{A\psi1}} \\[5mm] I_{\max B\phi1} = \dfrac{M\phi1 \cdot \left(\left(\left(k_{AB\phi1} - k_{A\psi2}\right) \cdot k_{B\psi1} + k_{AB\phi} \cdot k_{A\psi2}\right) \cdot M\phi - k_{AB\phi1} \cdot k_{A\psi1} \cdot I_{\max A\phi}\right)}{\left(k_{AB\phi} \cdot M\phi - k_{A\psi1} \cdot I_{\max A\phi}\right) \cdot k_{B\psi2}} \end{cases}$$

$$\tag{8.63}$$

To find the value of possible maximum shifts, we use the proportionality coefficients (PC_A and PC_B) of formulas (8.24) and (8.25).

$$I_{\max B\phi1} = PC_B \cdot I_{\max B\phi} = PC_B \cdot M\phi \tag{8.64}$$

$$I_{\max A\phi} = \frac{I_{\max A\phi1}}{PC_A} = \frac{M\phi1}{PC_A} \tag{8.65}$$

We use in formulas (8.60), (8.61), (8.64), and (8.65), proportionality coefficients (PC_A and PC_B), which are calculated on the basis of the parameters of the selected standards for the selected directions ($A\psi1$ and $B\psi1$, $A\psi2$ and $B\psi2$). When the values of the parameters $I_{\max A\phi}$, $I_{\max A\phi1}$, $I_{\max B\phi}$, and $I_{\max B\phi1}$ are known, we check the proportionality coefficients, calculating these coefficients based on these parameters. If equations (8.24) and (8.25) are truth, then we proceed to a detailed comparison of scaled *FAIs*.

Thus, the process of determining function pairs for detailed comparison will be as follows.

For each selected pair of reference functions for each of the nine variants of the maximum shifts determine the size of the possible real maximum shifts ($I_{\max A\phi}$, $I_{\max B\phi}$, $I_{\max A\phi1}$, $I_{\max B\phi1}$). Proportionality coefficients can be used for this purpose. Using formulas 8.34 and 8.35, we determine the values of the real areas of both objects (S_{0A} and S_{0B}) for each of the options. Substitute them into formula 8.4 to calculate the total area S_{0AB}. If for any of the variants it is equal to the initial value of the cyclic shift function $FAI_{AB}(0)$ for any direction, then the current pair of reference functions of the area of intersection can be transferred for further comparison at a detailed stage with the corresponding sections of really obtained total cyclic FAIs (FAI_{AB}) for directions φ and $\varphi 1$. If at the detailed stage of comparison the values of the corresponding functions coincide with the given accuracy, it can be concluded that the objects of the scene are identical in shape to the selected reference objects. Using the scaling factors ($SC_{A\psi 1}$ and $SC_{B\psi 1}$), we can determine their real parameters. This data will allow the intelligent system to fill the map of the surrounding space with models of real objects.

It should be noted that all calculations in this section were performed for the options of mutual location of objects 1 and 2 (Figure 8.1a and b). The intersection of objects 3 is characterized by the absence of directions that are not included in the intersection sectors. In this case, the intersection sector is equal to 2π for any type of shift (cyclic and non-cyclic). This greatly complicates the analysis of general cyclic functions to obtain the parameters of the scene objects.

8.4 The Order for Determining the Basic Parameters of the Scene

To determine the basic parameters of the elements of the scene, we must perform the following sequence of actions.

1. Obtain a set of cyclic functions of the area of intersection (FAI_{AB}) for the range of directions from 0 to π.

2. Determine the range of directions that are not part of the intersection sector. The total cyclic functions of the area of intersection that do not fall within this range in the area corresponding to the cycle length ($lenth_{\varphi}$) will have only one burst of values.

3. Use the offset band parameters to determine the parameters of the circumscribed rectangles for the scene.

4. For the selected direction of shift, which is not part of the intersection sector, determine the values of the corrective parameters. One of them is equal to the distance between the objects of the scene in the direction perpendicular to the selected one.

5. To determine the shape and quantitative parameters of the objects of the scene from the range of directions that are not part of the intersection sector, choose two directions that are not equal to each other ($\phi \neq \phi 1$).

6. For both selected directions, the parameters of the initial area ($S_{0AB} = FAI_{AB}(0)$) and the maximum shift ($\max(I_{maxA}, I_{maxB})$) are determined.

7. In the reference database, the parameters of a certain object (A) for some direction $A\psi1$ are selected for verification. The parameters of the same object for direction $A\psi2$ are also selected. If for all possible directions of the standard A the check is carried out, we pass to another standard A for which we choose functions for the directions $A\psi1$ and $A\psi2$

$$\left(\phi1-\phi = A\psi2 - A\psi1 = B\psi2 - B\psi1\right).$$

8. In the reference database, the parameters of another object (B) are selected to check for some direction $B\psi1$. This object can be identical to reference A. The parameters of the same object for direction $B\psi2$ are also selected. If all available reference directions $B\psi1$ of this object are checked, and there was no transition to a detailed stage of comparison, we pass to other object B. If all available reference objects B are checked and there was no transition to a detailed stage of comparison, then go to paragraph 7 and choose another direction $A\psi1$ or the next standard A.

9. At the fast stage of the comparison, the possible real parameters of the research objects are determined ($I_{maxA\varphi}$ and $I_{maxB\varphi}$, $I_{maxA\varphi1}$ and $I_{maxB\varphi1}$, S_{0A} and S_{0B}). The truth of equations on the basis of integral coefficients is checked. If this checking is passed, then proceed to the detailed stage (comparison of scalable functions of the area of intersection). Otherwise we pass to item 8 and we choose other direction $B\psi1$ or the following standard B.

10. At the detailed stage of the study, the values of the corresponding sections of the cyclic FAIs ($FAI_{AB\varphi}$ and $FAI_{AB\varphi1}$) for the directions φ and $\varphi1$ are compared with the sums of the scaled FAIs of the selected standards A and B for the directions $A\psi1$ and $B\psi1$, $A\psi2$ and $B\psi2$. Scaling is performed using scaling factors ($SC_{A\psi1}$ and $SC_{B\psi1}$). If the coincidence in all cases did not occur, then go to paragraph 8 and choose another direction $B\psi1$ or the next standard B.

11. If the comparisons for both fast and detailed stages are successful, then we will consider the current standards A and B with orientation with the corresponding directions of spatial orientation $A\psi1$ and $B\psi1$ identical to the objects of the scene.

12. If the results of the check in the reference database do not find a pair of objects, the scalable parameters of which for any direction do not allow to form the functions $FAI_{AB\varphi}$ and $FAI_{AB\varphi1}$, it is necessary to segment the image, and after that it is necessary to enter parameters of new objects to the reference database for train the system to use new images.

Since this procedure is intended for the analysis of scenes created by two objects, for scenes created by more objects, it would be advisable to divide the array of these elements into pairs. If the first element of each such pair is the second element of the previous pair, then in the process of analysis can be built a chain of elements, the relative position of which is known. Then, by determining by some other method the real location of any element of the chain, we can determine the real spatial position of all elements of the scene.

We can also segment images into certain identical elementary components. For several identical and identically directed regularly spaced elements, the formula for the interdependence of the integral coefficients and the initial areas will be as follows.

$$k_{AA...A\varphi} = k_{A\psi} \tag{8.66}$$

For each of the directions φ the spatial location of the elements of the reference image ψ (angle of inclination to the orthogonal axes) must coincide with a similar angle of inclination of the other elements of the scene.

$$S_{0AA...A} = N \cdot S_{0A} \tag{8.67}$$

where N is the total number of elements.

This property can be used to analyze objects that are a set formed of identical elements. However, the total cyclic function $FAI_{AA...A}$ will be very difficult to analyze if the distance between these elements is less than or equal to the linear size for each direction of shift (second variant of mutual location).

Thus, to solve both subtasks of analysis of scenes from two objects using methods of parallel shift technology, it is necessary to have a set of real cyclic functions of the area of intersection of the scene and reference data of objects that can be its elements.

9

Radon Transformation Technology on Cellular Automata with Hexagonal Coating

9.1 Introduction

As you know, the Fourier transform is successfully used in image recognition (Zalmanzon, 1989; Holland, 1966). But the processing of visual data based on the computationally intensive Fourier transform is rather slow and suitable for analyzing static images; therefore, it is unacceptable for systems operating in real time. The advent of the Fast Fourier Transform algorithm has stimulated interest in other types of spectral transformations. In the problems of filtering, compression, and extraction of informative features, such transformations as Hadamard, Hough, Haar, Hartley, and Radon are most widely used (Pratt, 2016; Solomon & Breckon, 2011; Nixon & Aguardo, 2002; Gonzalez & Woods, 2008; Belan & Motornyuk, 2013; Minichino & Howse, 2015; Jain, 1989; Deans & Roderick, 1983; Ell, Bihan, & Sangwine, 2014; Ryan, 2019; Richard & Duda, 1972; Zalmanzon, 1989). Despite their diversity, it turned out that most of the fast transformation algorithms have a similar structure and differ by no more than the values of the coefficients of the basic operations (Dorogov, 2011), which means they have the same disadvantages in the form of computational resources.

But among all these transformations there is a unique in its kind, radically different from the rest by the possibility of its hardware implementation on cellular automata – this is the Radon transform. Since cellular automata are by definition parallel structures, the implementation on this architecture provides an undeniable advantage in the speed of the algorithm in real time. This combination of technologies has allowed researchers to achieve high results in the field of intelligent data processing in relation to biometric identification (Motornyuk, Bilan, & Bilan, 2021; Bilan, Motornyuk, Bilan, & Galan, 2021d).

Such a multi-projection approach to the description of images is very effective, since it allowed the development of methods for processing and recognizing images of varying complexities (Motornyuk & Bilan, 2019a, 2019b;

Motornyuk, Bilan, & Bilan, 2021). They use methods to analyze the quantitative characteristics of the area (for parallel shear technology) and the quantitative characteristics of the projections obtained by the Radon transform. At the same time, the hexagonal coverage allows increasing the number of formed projections of the Radon transformation to six.

The originality in these technologies is that they collect information about the projected image in the field of view, which does not exceed 180°. Approximately the same sector of the human fundus is covered by the retina, the receptors of which are located on the semicircle of the fundus. This suggests that the visual channel also analyzes visual information projected for a sector of coverage not exceeding 180°.

The quantitative values of the characteristic features obtained as a result of the application of these methods are unique for each image. As a rule, significant differences are obtained by using multiple projections. If several projections are not enough, then additional shifts or Radon projections are used, which give an accurate description of the image.

Technologies make it possible to successfully deal with image noise, take into account affine transformations, and also successfully implement various operations of preliminary image processing. This is confirmed in various publications (Bilan, Motornyuk, & Bilan, 2014a). The constant use of such technologies reveals all new opportunities and advantages in various aspects of information processing.

9.2 Images Description Technology Based on Cellular Automata with Hexagonal Coated Radon Transform

An analysis of the features of three types of coverage used to construct cellular automata showed that to solve a number of image processing problems and, in particular, to effectively implement the Radon transform, it is most expedient to use a hexagonal coverage, since it does not have ambiguities in determining neighboring cells. In addition, it is the most natural and is often found in both living and inanimate nature (Figure 9.1).

In addition, confirmation of our conclusions is available in the scientific literature describing studies on the comparison of hexagonal and orthogonal coverage in terms of their effectiveness in image processing operations (Dudgeon & Mersereau, 1984; Golay, 1969; Staunton, 1989a, 1989b; Mersereau, 1979). From the analysis of the studies carried out, conclusions follow about the main advantages of the hexagonal coating:

requires 13.4% fewer points to represent the same signal (Dudgeon & Mersereau, 1984; Mersereau, 1979; Woodward, 1984);

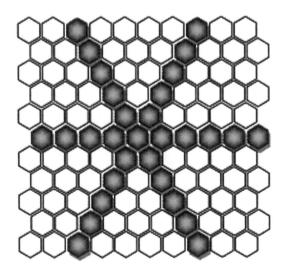

FIGURE 9.1
The main directions of hexagonal coverage.

simple and unambiguous definition of links (Golay, 1969);

more efficient local and global computing (Staunton, 1989b; Mersereau, 1979);

simplicity in operator development (Staunton, 1989a);

But the main advantage of the hexagonal coverage is that, due to the geometric features of the lattice, three main directions 0°, 60°, and 120° are distinguished on it (Figure 9.2), which exactly coincide with the minimum set of directions for constructing the Radon transform (RT) projections required to restore the original image. And this not only provides simplicity of hardware implementation, but also proves that a set of three such projections can be considered a characteristic feature for image identification.

The physical meaning of the RT for a two-dimensional binary image is to find the sum of the pixels that form a given image along a straight line in a certain direction (along which the transform is carried out). The results of such transformations will be arrays of values.

In computed tomography, the Radon transformation is performed along the entire set of directions of the circle (2π), since the main task in medicine is the restoration of the initial image. But for the extraction of characteristic features in digital image processing, such a solution is rather redundant; therefore, in most cases, it is enough to build projections only along a small set of certain directions. Moreover, it is enough to consider only a sector of 180° (π), since for any angle α its resulting array will be a mirror image of the array for the angle $\alpha + 180°$. This is clearly seen in Figure 9.2: for angles 0° and

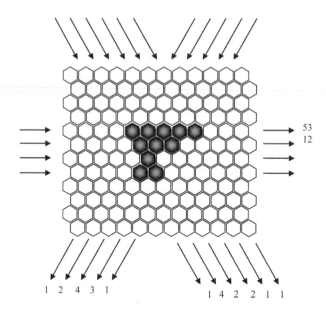

FIGURE 9.2
Construction of projections of the Radon transform in the directions 0°, 60°, and 120°.

180° we will get "mirror" results, as well as for angles 60° and 240°, 120° and 300° (the direction arrows will point in the opposite direction).

In total, on the hexagonal covering, six "innate" natural directions can be distinguished, formed by the geometric features of this structure. Figure 9.1 shows a fragment of the cellular environment, where arrows indicate directions corresponding to the main directions with angles 0°, 60°, and 120°. Figure 9.3 shows a fragment of the cell environment with additional directions with angles of 30°, 90°, and 150°.

We use this feature of the hexagonal lattice to construct cellular automata that perform the RT in hardware with an angle step of 30° (6 directions) or 60° (3 directions). Moreover, the number of directions can be selected depending on the nature and type of the range of input images, the characteristic features of which must be extracted.

A cellular automata is a matrix that in general appearance repeats a single pixel – a regular hexagon, and its size can be arbitrary depending on the required number of cells (processing elements). Thus, the maximum number of cells in a row of the main direction will be determined by the formula:

$$Z_{\max} = 2 \cdot n - 1 = k \tag{9.1}$$

The minimum number of cells in a row of the main direction:

$$Z_{\min} = n \tag{9.2}$$

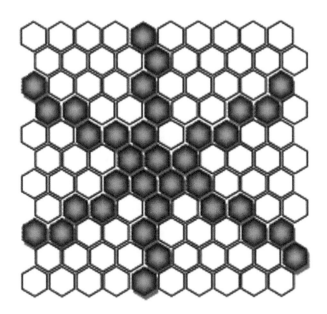

FIGURE 9.3
Additional directions of hexagonal coverage.

Maximum number of cells in a row of additional direction:

$$Z'_{max} = Z_{max} = 2 \cdot n - 1 = k \tag{9.3}$$

Minimum number of cells in a row of additional direction:

$$Z'_{min} = 1 \tag{9.4}$$

where n – is the number of cells that make up one side of the cellular automaton and k – maximum line length.

Taking into account the specific values of the angles for constructing projections, the well-known RT formula

$$\check{f}(p,\varphi) = Rf = \int_{-\infty}^{\infty} f(p\cos\varphi - s\sin\varphi, p\sin\varphi + s\cos\varphi)ds \tag{9.5}$$

can be written for each direction of a cellular automata with hexagonal coated:

for direction 0° ($R_0 f$):

$$\check{f}_0(p,0) = R_0 f = \int_{-\infty}^{\infty} f(p,s)ds \tag{9.6}$$

for direction 30° ($R_{30}f$):

$$\overset{\vee}{f}_{30}(p,30) = R_{30}f = \int\limits_{-\infty}^{\infty} f\left(\frac{\sqrt{3}}{2}p - \frac{1}{2}s, \frac{1}{2}p + \frac{\sqrt{3}}{2}s\right)ds \qquad (9.7)$$

for direction 60° ($R_{60}f$):

$$\overset{\vee}{f}_{60}(p,60) = R_{60}f = \int\limits_{-\infty}^{\infty} f\left(\frac{1}{2}p - \frac{\sqrt{3}}{2}s, \frac{\sqrt{3}}{2}p + \frac{1}{2}s\right)ds \qquad (9.8)$$

for direction 90° ($R_{90}f$):

$$\overset{\vee}{f}_{90}(p,90) = R_{90}f = \int\limits_{-\infty}^{\infty} f(s,p)ds \qquad (9.9)$$

for direction 120° ($R_{120}f$):

$$\overset{\vee}{f}_{120}(p,120) = R_{120}f = \int\limits_{-\infty}^{\infty} f\left(-\frac{1}{2}p - \frac{\sqrt{3}}{2}s, \frac{\sqrt{3}}{2}p - \frac{1}{2}s\right)ds \qquad (9.10)$$

for direction 150° ($R_{150}f$):

$$\overset{\vee}{f}_{150}(p,150) = R_{150}f = \int\limits_{-\infty}^{\infty} f\left(-\frac{\sqrt{3}}{2}p - \frac{1}{2}s, \frac{1}{2}p - \frac{\sqrt{3}}{2}s\right)ds \qquad (9.11)$$

where

ds – increase in length along the line of construction of the projection (9.5)–(9.11),

p – distance from the origin to this line.

Let Q be the set of main and additional directions of the hexagonal covering: $Q = \{0°, 30°, 60°, 90°, 120°, 150°\}$. If every i-th line ($i = \{1, 2, ..., k\}$, where k – maximum line length) in the q direction is represented as a set of pixels Lqi, and such that Lqi = {x1, ..., xk}, then we can write the formula for the sum of the brightness values of all pixels of this line:

$$S_{qi} = \sum_{j=1}^{Z_i} x_j, \qquad (9.12)$$

where Zi – the maximum number of pixels of the i-th line lying within the boundaries determined by the formulas (9.1)–(9.4).

Then, taking into account (9.12), the projection of the RT for the direction q can be represented as an ordered set Rq consisting of all the sums Sqi of the given direction:

$$R_q = \left\{ S_{q1}, S_{q2}, \ldots, S_{qk} \right\}. \tag{9.13}$$

Since formulas (9.6)–(9.11) imply summation, then for the hardware implementation of the RT on a cellular automaton, the entire image is shifted outside the environment, while counting on the extreme processing elements the number of image pixels passed through them. A total of six offsets are performed, three of them being performed along the main directions (Figure 9.4 top row), and three along additional directions (Figure 9.4 bottom row). To perform the Radon transformation (Figure 9.4):

for angle 0° – shift the image to the left, each processor element (x0) takes the value of its right neighbor (x4);

for an angle of 60° – shift the image to the left and down, each processor element (x0) takes the value of its upper right neighbor (x2);

for an angle of 120° – we shift the image to the right and down, each processor element (x0) takes the value of its upper left neighbor (x1);

for an angle of 30° – we perform image shift left and down, each processor element (x0) alternately takes the values of its upper right (x2) and right (x4) neighbors;

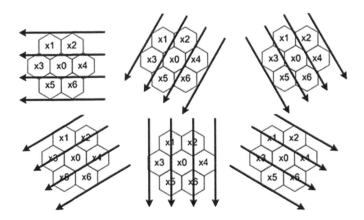

FIGURE 9.4
Directions of shifts in the processor element.

for an angle of 90° – we perform a downward shift of the image, each processor element (x0) takes the values of its upper right (x2) and upper left (x1) neighbors in turn;

for an angle of 150° – we perform a shift of the image to the right and down, each processor element (x0) alternately takes the values of its left (x3) and upper left (x1) neighbors;

Thus, shifts in additional directions are technically performed as alternate single shifts in two adjacent main directions. Having combined the location and size of all six projections, it is sufficient to place the summing elements not along the entire perimeter of the matrix, but only in accordance with the diagram shown in Figure 9.5. A general view of a cellular automaton with a hexagonal covering, which constructs six projections of the Radon transform, is shown in Figure 9.5, where the active zone for drawing the image is shown in gray, and the extreme processing elements performing the addition operations are indicated in dark.

The total number of processing elements (p) in such a cellular automaton depends on n – the number of cells that make up one side of it, and is determined by the formula:

$$p = 3n \cdot (n-1) + 1 \qquad (9.14)$$

The total number of processor elements of the active area (Figure 9.5) is determined by the formula derived empirically:

$$a = n \cdot (n-1) + b \qquad (9.15)$$

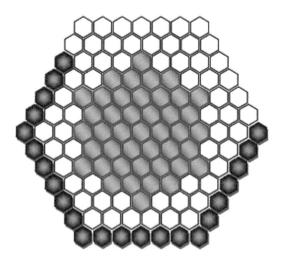

FIGURE 9.5
General view of cellular automata performing the Radon transform.

where b – coefficient determined as the remainder from dividing $n/3$: $b = -1$, if the remainder = 2, and $b = 1$ in all other cases.

Estimation of the speed of the cellular automaton performing the Radon transform is done. Let's calculate the speed of the developed cellular automaton with an image size of 2048 pixels wide. This image should fit into the active area (gray cells in Figure 9.5). Then the side of the cell structure will also consist of n = 2048 pixels. The approximate number of auxiliary points (white cells in Figure 9.5) in the main directions is sin(30)*2048 = 1024. We obtain the maximum displacement distance 2048 + 1024 = 3072. Approximately the same number of shifts are in additional directions. In total, 6 projections of the Radon transformation are allocated, that is, 6 operations take place in 3072 time steps: 6*3072 = 18,432. That is, at a cellular automaton frequency of at least 1 MHz, the construction of six projections of the Radon transform for an image of the order of 4 megapixels will be performed in no more than 0.0184 s. And this is approximately 54 frames/s, which is quite enough for the device to work in real time.

10

Biometric Identification Methods Based on the Geometry of the Auricle Based on Parallel Shift Technology and Radon Transformation

10.1 Technologies for Biometric Identification

For biometric identification of a person, various technologies are used, which are based on various theories related to signal processing, transmission, and analysis of data. Various technologies are used to analyze different biometric characteristics. As mentioned earlier, the most commonly used are image processing and recognition technologies. These technologies are implemented on various theories, which include many theories of image transformation: Fourier transform, Hough, Haar, wavelets, artificial neural networks, etc. (Pratt, 2016; Solomon & Breckon, 2011; Nixon & Aguardo, 2002; Gonzalez & Woods, 2008; Belan & Motornyuk, 2013; Minichino & Howse, 2015; Jain, 1989; Deans & Roderick, 1983; Ell, Bihan, & Sangwine, 2014; Ryan, 2019; Richard & Duda, 1972; Zalmanzon, 1989).

However, each biometric characteristic differs from the others in that different characteristic features are used to describe it. This situation forces developers to develop identification methods that aim at analyzing only one biometric characteristic. Moreover, for each biometric characteristic, many biometric identification methods have been developed, which, as a rule, analyze the same set of characteristic features.

Along with these methods, methods are also used that rely on only one basic, inherent in the method, values. The generated values have a unique meaning for each individual. In this case, the generated values can be represented in various forms (number series, graphical form and table). In a computing system, the uniqueness of a biometric characteristic is most often represented by a unique ordered set of numbers. Such technologies include: parallel shift technology (PST) (Belan & Yuzhakov, 2013a, 2013b; Bilan, Yuzhakov, & Bilan, 2014b; Yuzhakov, 2019), identification based on Radon transformation

(Motornyuk & Bilan, 2019a, 2019b; Bilan, Motornyuk, & Bilan, 2014; Bilan, Motornyuk, Bilan, & Galan, 2021d), identification based on cellular automata (Bilan, Bilan, & Motornyuk, 2020; Bilan, 2017), and other technologies.

These technologies are aimed at image processing and for this they implement permanent algorithms for any form of homogeneous data. However, biometric characteristics are represented by a combination of different components. For example, the geometric shape of a human palm can be described by more than 70 characteristics that quantitatively describe its separate unique segment (finger length, palm width, etc.). Each such segment of biometric characteristics may require unique processing methods.

There are biometric characteristics that are tied to only one device and use unique identification methods (Bilan, Elhoseny, & Hemanth, 2021c). These biometric characteristics include, for example, the dynamics of working on the keyboard (Bilan, Bilan, & Bilan, 2021a).

How else can to identify or authenticate a person?

The development of modern network and information technologies has given impetus to the technology of human identification, based on the creation of their own identifier, which is recorded in the corresponding database of identifiers. These identifiers include: a numeric code, a key phrase, a self-created image, a self-created sound signal, etc. This technology involves hiding (encrypting) an identifier that is inaccessible to everyone. A person in this case can be identified only if he himself wants it or agrees to be identified.

The latter method characterizes individual, psychological, and creative properties when forming a password identifier, and is also more democratic. In a smart city environment, this method is very widely used, since it is characterized by complete automation of all processes, which requires remote identification of a person to manage multiple objects and carry out various operations in many areas (banks, transport, etc.).

The chapter describes the methods of biometric identification, which were developed and researched by the authors based on their own copyright technologies. Mainly, methods of biometric identification using biometric characteristics recorded by a video camera are considered. It describes technologies for processing images of biometric characteristics, as well as technologies for multimodal biometric identification of a person.

10.2 Biometric Identification Methods Based on the Analysis of the Auricle

The geometric shape of the human ear is one of the biometric characteristics of a person, which is often used for biometric identification. The geometric shape of the human auricle has a simple structure that can be easily described using modern computational image processing tools. At the same time, the shape of the auricle in details for each person differs from the shapes of the auricles of other people, which rightfully puts it on a par with the main bio-metric characteristics of a person.

At the same time, the uniqueness of the auricle at this point in time has no clear evidence. Therefore, ears are used as an additional characteristic in the construction of multimodal intelligent systems.

Human ears are characterized by a number of features compared to other visual biometric characteristics of a person. These include (Motornyuk, Bilan, & Bilan, 2021):

- All elements of the auricle are displayed with similar color characteristics;
- Rigid and unchanged arrangement of all elements of the auricle;
- Small size of the occupied area of the image of the auricle;
- Independent of the location of the sensor.

These characteristics are among the advantages that encourage developers to use the image of the auricle as the main biometric characteristic. The first characteristic makes it possible to simplify the methods for selecting the image of the auricle in the image of the human head, if the ear is not covered by other elements (hair, headdress, etc.). The second feature characterizes the invariability of the shape of the image of the auricle, which does not require additional means for rigid fixation of its elements. The third feature does not require additional scaling of images, and the fourth feature allows you to capture the image of the auricle at different distances, and also does not require a special positioning of the human head.

The main problem of biometric identification from the image of the auricle is the selection of this image against the background of the rest of the human head and neck. However, in this regard, there are a large number of methods that, under certain conditions, give good results.

Many works have been devoted to the implementation of biometric identification systems (Bilan, Elhoseny, & Hemanth, 2021c; Boulgouris, Plataniotis, & Micheli-Tzanakou, 2009; Fairhurst, 2019). All these works determine the geometric shape of the auricle, which is described by a vector of characteristic features. The generated vector of characteristic features is compared with the reference vectors stored in the reference memory.

All the developed methods are characterized by methods for extracting the auricle. The methods for extracting and fixing the auricle depend on the method of capturing the image of the ear:

- a clear selection of only the auricle;
- fuzzy selection of the auricle in the image;
- close scanning of the auricle.

FIGURE 10.1
Examples of images of the auricle, with a clear selection.

The simplest method is realized with preliminary selection of only the image of the auricle. This selection is realized by cutting off the images of the rest of the body from the image of the ear and forming a background (for example, a white background). Such a background can be formed with the help of special plates or paper with a hole in which the auricle is placed, and a plate or paper tightly adhering to the head at the back. This method creates certain inconveniences for the identified person. The color of the plate should not give color reflections when fixing the image of the auricle. Examples of clear image selection are shown in Figure 10.1.

Almost all methods for analyzing the image of the auricle are based on the selection of control points in the image and their further analysis of their location for each image. This work uses the method described in the work by Motornyuk, Bilan, and Bilan (2021).

To implement the method, perform the following actions.

1. With the help of special tools, the image of the auricle is selected on a uniform background.
2. Binarize the selected image. In this case, binarization can be carried out for different color and brightness thresholds.
3. A rectangular field is formed, which describes the binary image of the auricle at the extreme points.
4. A binary image of the auricle is analyzed and a vector of characteristic features is formed, which contains an ordered set of quantitative data.

All biometric identification methods are implemented by these operations. Distinguish each method from others by detailed implementation of the first and fourth steps. If a clear selection of the image of the auricle is considered, then the first step of the method is the same for all other methods.

This paper uses a method based on the method described in Motornyuk, Bilan, and Bilan (2021). This method is characterized by the following.

After the formation of the circumscribing rectangle, a horizontal line is drawn through the widest area of the image of the auricle, which divides the circumscribing rectangle into two rectangular areas. The resulting lower rectangular area is divided by a horizontal line into two equal parts. A vertical line is drawn through the horizontal line with the most black pixels.

After dividing the area of the auricle image into six subareas, the area of the image of the entire auricle and the area of each of the six selected subareas are calculated. The ratio of the areas of small subregions to the total area of the image is calculated, and also the relationship between the lengths of the additional and main horizontal lines is calculated.

In addition, Radon projections are calculated for the general image and the relationship between them is determined. From the obtained quantitative values, a vector of characteristic features is formed, which is compared with the reference vectors stored in the reference database. Based on the comparison results, a decision is made on the identification result.

A graphic display of the result of splitting the binary image of the auricle is shown in Figure 10.2.

With a clear selection of the image of the auricle, the geometric shape of the binary image practically does not change. This method requires a solid image filled with black pixels inside the outline. Otherwise, there is a possibility of false calculation of the main vertical and horizontal lines. In this

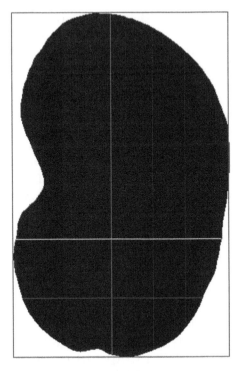

FIGURE 10.2
Dividing the binary image of the human auricle into six subareas.

FIGURE 10.3
Example of the image of the curves of the outer edges of the contour for each selected subarea of the image.

case, the binarization threshold should be clearly selected, which excludes the appearance of white background pixels.

Also, subregions of different shapes can have the same area, which can lead to false identification. The main distortions are made by the extreme outlines of the auricle. Therefore, it is necessary to carry out additional analysis of the curves of the outer contour of each subregion. An example of such curves for each separate area is shown in Figure 10.3.

Cellular automata technologies were used to edge detection (Bilan, Bilan, & Motornyuk, 2020; Bilan, 2017). In this example, the outline was selected based on the von Neumann neighborhood. The resulting contour can be analyzed using discretization of the imaging surface. If you look closely at the details, you can see the staircase effect (Figure 10.4). As a result of the staircase effect, the number of cells (pixels) encoding 1 (belonging to the contour) is counted vertically and horizontally. The resulting sequences characterize the curves of the outer contour of each selected part of the image.

The alternation of numbers indicates the characteristic features of the curve shape. For the example shown in Figures 10.3 and 10.4, the sequence of numbers formed along the horizontal axis is as follows:

- for the first part of the auricle image

    0,0,0,0,3,7,9,3,3,2,2,21,11,11,11,7,8,6,6,5,7,6,7,6,4,6,5,7,7,5,4,6,4,5,5,4,5,6,6, 7,8,13,1,2,2,1,2,1,1,1,1,1,2,1,1,1,1,1,1,1,1,1,1,1,1,1,1,1,1,1,1,1,1,1,1,1,1,1,1,1, 1,1,1,1,1,1,1,1,1,1,1,1,1,1,1,1,1,1,1,1,1,1,1,1,1,1,1,1,1,1,1,1,1,1,1,1,1,1,1,1,1, 1,1,1,1,1,1

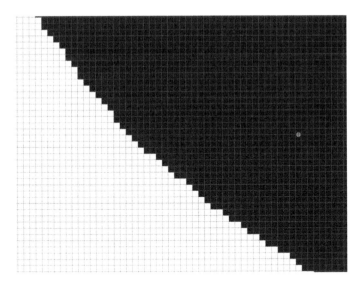

FIGURE 10.4
Image of aliasing on a discrete orthogonal surface.

- for the third part of the auricle image

0,0,0,0,0,0,0,0,0,0,0,0,0,2,3,3,3,3,3,3,4,2,3,2,2,2,3,2,3,2,2,2,2,2,3,2,2,2,2,2,2,2,2
,2,2,2,2,2,2,2,2,2,2,2,2,2,2,2,2,2,2,22,2,2,2,2,2,2,2,2,2,2,2,2,2,2,2,2,2,2,2,2,2,2,2
,2,2,2,2,2,2,2,2,2,2,2,2,2,2,2,2,2,2,2,2,2,2,2,2,2,2,2,2,2,2,2,2,2,2,2,2,2,2,2,2,2,2

The presence of large numbers at each position indicates an approximately vertical position of the curve section. By the ratios of large and small numbers in the sequence, one can judge the behavior of the curve.

You can also analyze the curves and the obtained Radon projections (Motornyuk & Bilan, 2019a, 2019b; Bilan, Motornyuk, & Bilan, 2014a; Bilan, Motornyuk, Bilan, & Galan, 2021d). However, to describe the curve, it is necessary to analyze several Radon projections, which greatly complicates the analysis algorithms. Therefore it is better to use contour analysis.

In a smart city, it is not always possible to use special means for separating the image of the auricle, since this requires the consent of the person himself and also leads to additional time costs. This situation does not fully automate the identification process.

Solving the problem of complete automation of the process of human identification based on the image of the auricle requires additional complex algorithms aimed at automatic selection of the image of the auricle against the background of images of the rest of the body adjacent to the image of the human ear. A large number of auricle images freely available on the Internet is presented by Esther Gonzales. Examples of such images are shown in Figure 10.5.

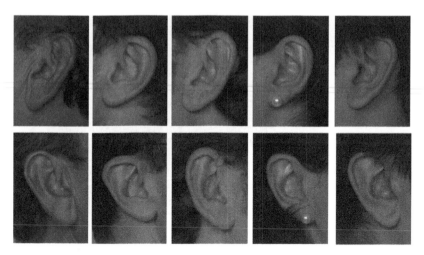

FIGURE 10.5
Images of the auricles from the Esther Gonzales database.

Images are characterized by a wide range of brightness and color parameters. This requires the use of complex algorithms for detailed image analysis.

To solve this problem, an algorithm is used, which is a modification of the method described in the work by Motornyuk, Bilan, and Bilan (2021). According to the algorithm, the image from the video sensor is formed in the computing system as a two-dimensional array of numbers. Each number displays the code of the corresponding pixel that encodes the color and brightness of the pixel.

The algorithm uses the fact that the system has a predefined structure of the image, and the system "knows" in advance the codes of the pixels that belong to the image of the auricle. The system "does not know" the exact location of these pixels in the image. It is also taken into account that hair can be located on top and on both sides in the image, which has a characteristic image structure. In addition, pixels are located below the auricle that belong to the image of the body not covered with hair.

Analysis of the pinna images showed that the pixels belonging to the pinna are darker than the pixels belonging to other parts of the body, but not the hair.

To accomplish these positions, multiple images have been analyzed. As a result, the codes of the pixels belonging to the image of the auricle were determined. Moreover, the code is not the same, since people have different skin colors. To determine such a code in each specific case, the image containing the auricle is divided into areas in advance of certain sizes. Each area covers a certain number of pixels in the image. The areas intersect with each other and can be shifted by one or more pixels (Figure 10.6).

Figure 10.6 shows the opening image in the first row. The second row shows selected areas with a size of 220 × 350 and a threshold value equal

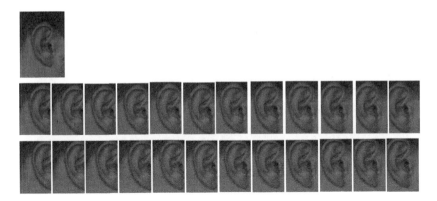

FIGURE 10.6
Example of dividing an image into averages.

to 3,300,000. A confidence interval of 500 was also used. In the third row, a threshold value of 3,500,000 was used.

As can be seen in the second row 6, 7, 8 and 9, the images clearly and completely select the image of the auricle. At a threshold of 3,500,000, the upper part of the auricle does not fall slightly into the limiting window, which can lead to false identification results.

The average is calculated for each area. A field is defined that indicates the approximate value of the pixel code. If the area of the field is less than the area of the image of the auricle, then a whole group of fields with approximately equal mean values is determined. At the same time, they are located in the image area of the auricle, in fact, they outline it.

After determining the approximate average value of the code of the pixels belonging to the image, the selection area is increased. The size and shape of the area of the final enclosing area are calculated based on the analysis of the location of the smaller areas. Moreover, there can be several such areas (Figure 10.7). In this case, all areas that show an approximate average value are processed.

The images of auricles in the selected areas are processed so that only those pixels whose values fall within the specified range are selected.

The following formula is implemented

$$G(t+1) = \begin{cases} G(t), \text{if } G_{mean} - F < G(t) < G_{mean} + F \\ 0, \text{in othe case} \end{cases} \tag{10.1}$$

where

$G(t)$ – the value of the pixel code at time t before extraction;

G_{mean} – taken average value of the code of the pixel belonging to the auricle image;

F – value indicating the size of the confidence interval.

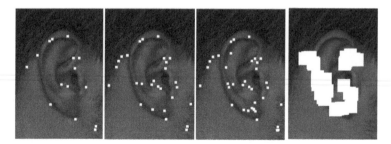

FIGURE 10.7
Process of extracting a pinna image using enclosing areas with approximate code means.

As a result of applying formula (10.1), pixels are allocated that fall within the accepted confidence interval; the remaining pixels are encoded as 0 and have black color. Figure 10.7 shows that with an increase in the confidence interval, the number of areas with the selected mean value increases. The first three figures use a 10 × 10 area and confidence intervals 200, 500, and 700, respectively. The fourth figure uses a 50 × 50 area and a 500 confidence interval.

As a result of the selection, the image contains selected pixels that do not belong to the area of the auricle. Almost all of these pixels are located below the auricle, on the left (if the auricle is turned to the left in the image) or on the right (if the auricle is turned to the right). The system is pre-trained to find such areas in the image and aims to remove them.

To remove unnecessary selected pixels, the Radon transform is used, according to which six Radon projections are formed due to the hexagonal coverage of the image (Motornyuk & Bilan, 2019a, 2019b; Bilan, Motornyuk, & Bilan, 2014a; Bilan, Motornyuk, Bilan, & Galan, 2021d). Initially, reference projections for the auricles with different directions are formed.

The obtained projections at the input of the system are compared with the reference projections, as a result of which extra areas on the input projections are determined. These extra areas of projections indicate the sector of the input image, in which the extra pixels are located. Such sectors are shown in Figure 10.8.

Analysis of the existing reference Radon projections and the Radon projections of the input image indicates redundant areas containing non-zero values. These areas go beyond the limits of the reference projections and therefore are removed by the system. The sequence of removing extra black pixels is shown in Figure 10.9.

As a result of the analysis, an approximate image of the auricle is obtained. Numerical sequences of each Radon projection are formed for each auricle and compared with the reference sequences of Radon projections. Comparisons are made using the selected confidence intervals. The thresholds of binarization of images are also taken into account, at which the Radon projection is calculated. In addition, the distance from the sensor to the person is taken

FIGURE 10.8
Sectorized images from six Radon projections.

into account, as well as the angle of the video camera in relation to the person's head.

It was experimentally found that changing the binarization threshold does not lead to significant changes in the forms of Radon projections. Figure 10.10 shows two Radon projections obtained in the 0° direction and formed at binarization thresholds of 60%, 70%, and 80%.

Figure 10.10 shows that there are no significant changes in the shape of the curves. There are changes in the range of coordinates from 153 to 177. However, they do not affect the overall percentage of coincidences within the accepted confidence interval.

The proposed methods require large amounts of memory for storing reference data. Especially if the system does not require a rigid position of the head identified to the image sensors. In this case, additional reference sequences are stored for each person, formed at different angles of the image sensors. Large volumes of reference sequences entail a large amount of time to find the nearest reference. However, identification accuracy is significantly increased.

The described methods were evaluated by determining the FFR and FAR values for each confidence interval. The FFR and FAR curves allowed us to determine the optimal threshold average percentage of matches, which

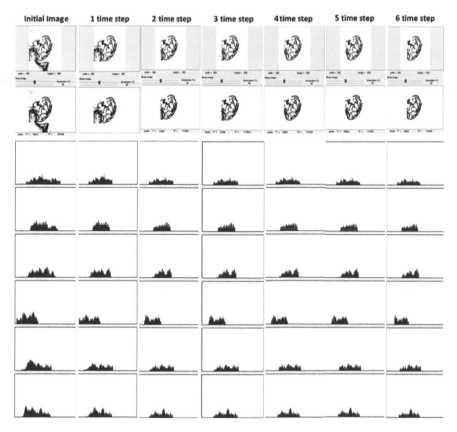

FIGURE 10.9
Example of sequential removal of redundant pixels by analyzing Radon projections.

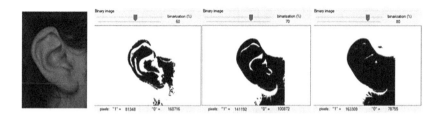

FIGURE 10.10
Two Radon projections for the pinna image taken in the 0° direction and at binarization thresholds of 60%, 70%, and 80%.

corresponds to 25%, which improves the indicator by 2% in comparison with the closest analogues. It was also found that for biometric identification of a person from the image of the auricle, there must be at least ten reference sequences in the database for each location of the image sensor.

10.3 Determination of the Area of Recognition in Biometric Identification by the Shape of the Ear Based on the Parallel Shift Technology

Certain static methods of biometric identification are based on the analysis of the shape of some parts of the human body. One such feature is the shape of the person's auricle. The area of this part of the body is not difficult to determine, so the use of PST can be productive.

One of the main tasks in the process of biometric identification by the shape of the auricle is to determine the area of recognition. The brightness of the ear image is close to the brightness of the skin of the face and neck fields. The hair that border this part of the body may have a brightness different from it. There is also an option when hairs are missing. Headphones, spectacle frames, hearing aids, and other devices can interfere with recognition. There is a need to highlight a certain area on the image of the head, where the ears are depicted.

10.4 Determining the Cut-Off Points of the Ear Image

For clear identification by analyzing the shape of the ear, the person's head should be positioned in profile. The brightness of the skin of the ears and face are close. On the image of the head, it is necessary to determine a certain line that will separate the image of the ear from other parts of the head (Figure 10.11a). This line should pass through points that belong to both the head and the ear (Figure 10.11b). These points are the vertices of certain acute angles, which are part of the elements of the image of the head and differ in brightness from other areas (Figure 10.11c). These elements can be a shadow around the ear (due to its volume) or the edge of the auricle (Figure 10.11d).

Therefore, the first step in determining the location of the ear image should be to select the area of the shadow around the ear or the edge of the auricle. This can be done by analyzing the brightness of areas of the image. In order to obtain the coordinates of the points through which the line separating the image of the ear from the image of the head passes, it is necessary to use the

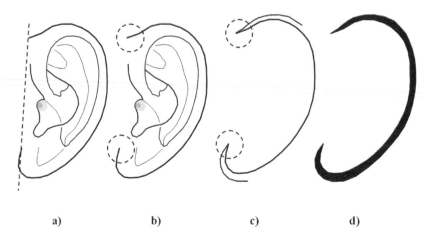

a)   b)   c)   d)

FIGURE 10.11
Determining the location of the line that separates the ear from other parts of the head.

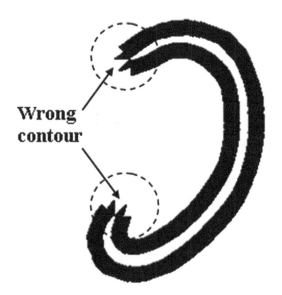

FIGURE 10.12
Incorrect contours of internal acute angles.

features of determining the contour of the image by the methods of PST. Let's call these points cut points. One of the disadvantages of contour selection using PST is that there is some distortion of the contour of the non-convex image in the presence of internal acute angles (Figure 10.12). Let's turn this disadvantage into an advantage.

FIGURE 10.13
Snowflakes noise due to contour selection.

This example shows a contour width of 15 pixels. When selecting the mentioned method of contours up to three units wide in images of any shape, this feature is not detected. The copy of the image is shifted when the contour is selected in eight directions. Usually in the real image there are noise inclusions. Because PST processes images for which an area needs to be calculated, noise should be considered to be individual elements of a small area or a line with a small width. Due to the presence of such noise inclusions in the process of allocating the circuit, there are distortions of another type. Let's call them "snowflake". The greater the width of the contour that stands out, the greater the rays of such "snowflakes" (Figure 10.13). This example shows "snowflakes" after selecting a contour with a width of 9 pixels.

Images of the edge of the auricle or the shadow around the ear have no internal sharp corners. In order to be able to get the wrong angles image, their original image must be inverted (Figure 10.14).

The inverse image has internal sharp corners. The contour with a width of pixels is selected. To select the wrong angles in the image, it is necessary to scan a circular area with a diameter $h - 1$. At each step, the internal filling of the scan area can be of four types (Figure 10.15a). The first type – the circular area consists of two parts: the background area and the contour. The second type – this area completely coincides with the contour section. This determines the choice of the diameter of the circle less than the width of the contour. The third type – the scanning area includes one section of the background and several sections of the contour. This option is possible when the original image is created by closely spaced lines. Detection of such elements can be useful in the analysis of papillary fingerprint lines. To detect wrong angles, it is especially important to determine the filling of the fourth type.

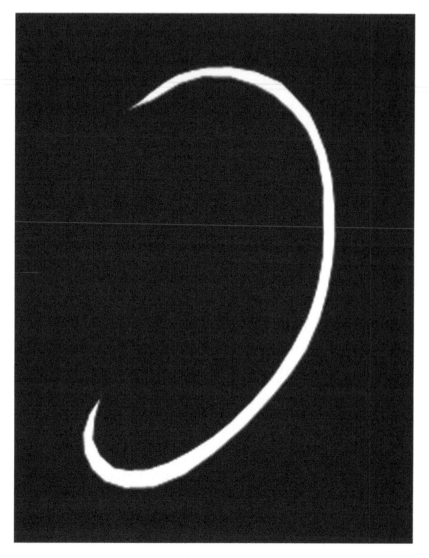

FIGURE 10.14
Inversion of the image of the shadow around the ear or the edge of the auricle.

This type is characterized by the presence of one section of the contour and two sections of the background. Moreover, if the width of the contour value is even, then the diameter of the circle is odd. Then, if the central pixel of the circular area is a contour section, and the two background areas have an area of close value, then this central pixel is a cut point (Figure 10.15b).

Since the contour width must be even and greater than 4 units, and increasing the contour width leads to a greater impact of noise inclusions, it is

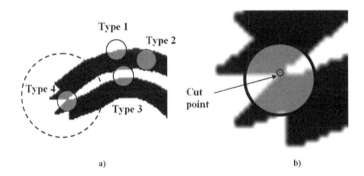

FIGURE 10.15
Determining the location of the cut point.

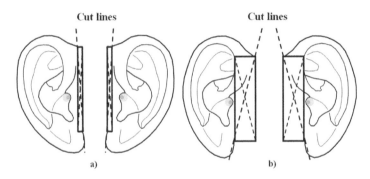

FIGURE 10.16
Examples of arrangement of cut rectangles.

optimal to use a contour width of 6 units to detect wrong angles. Then the diameter of the scan area will be 5 units. It should be noted that this method, like all methods of PST can be used for both raster and non-raster images. Therefore, values are given in units, not pixels. For digital images, pixels can be single elements.

A cut line passes through two cut points, which separates the image of the ears from the image of other parts of the face. Depending on the shape and location of the ears, there are different options for the location of cut lines. Parts of these lines coincide with the diagonals of certain cut rectangles external (Figure 10.16a) and internal (Figure 10.16b). One of the pairs of opposite vertices of these rectangles coincides with the cut points.

Due to the possible presence of noise inclusions in the image of the head when determining the cut line there is a need to verify the truth of the location of cut points. Several such points may be located in the image, some of which are wrong.

10.5 The Method of Dividing the Image into Individual Objects

The process of determining the cut line is checked by dividing the image into separate objects. Objects will be considered groups of pixels that border each other, have the same brightness (color), and their sum in area is greater than a certain maximum number of noise inclusions. At the beginning of the processing the image is created by elements that will be considered the background. There are a number of them. Let's call it the initial number of objects (N_{obj0}). Let's select the elements of the edge of the ear or the shadow around it in the image (Figure 10.12d). This group of items will not be considered an object. And let's draw a cut-off line from one edge of the receptor field to the other through the cut-off points (Figure 10.17a). We will not consider this line as an object either. The appearance of a cut line in the receptor field will increase the number of existing objects by at least two. One of the new objects is the area of the ear, and the number of other objects through which the cut line passes is doubled (Figure 10.17b). The total number of objects will be equal to N_{obj1}. For the given example, it will be the following.

$$N_{obj1} = 2 \cdot N_{obj0} + 1 = 3 \qquad (10.2)$$

Let's carry out certain manipulations over the cut line. An important factor in the case of a digital image for further action is the coherence of its elements. The distance of parallel shift of a cut line at the following steps depends on this indicator. Perform two parallel shifts to the right (Figure 10.17c) and to the left (Figure 10.17d). These actions change the number of objects that create the image ($N_{obj2right}$ i $N_{obj2left}$ respectively). The offset distance should be as small as possible so that the number of objects through which

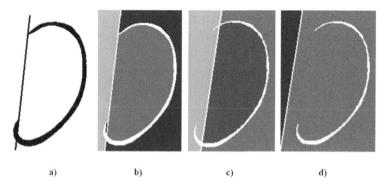

    a)   b)   c)   d)

FIGURE 10.17
Verification of the coordinates of the cut-off points.

the cut line passes does not change. With true cut-off coordinates, the number of objects changes as follows.

$$\begin{cases} N_{obj2right} = N_{obj1} \\ N_{obj2left} = N_{obj1} - 1 \end{cases} \tag{10.3}$$

In this case, the truth of the equations of the system 99 signals that the face of the person whose ear image is analyzed, turned to the left. If this system is transformed into the next,

$$\begin{cases} N_{obj2right} = N_{obj1} - 1 \\ N_{obj2left} = N_{obj1} \end{cases} \tag{10.4}$$

then the face of this person is turned to the right.

Thus, the correct position of the cut line increases the number of objects that create the image by at least two, and with minimal parallel shifts of the cut line, one of the systems of equations (10.3 or 10.4) is true. The cut line, together with the area formed by the edge of the ear (the shadow around it), delimits the area of the ear image. This area is analyzed in the process of biometric identification by the shape of the ear. The result of its search is the solution of the subtask of finding the area of the ear image.

If there are noises in the image of the head (hairstyle, hat, eyeglass brackets, hearing aid, headphones, etc.), which do not allow to detect the coordinates of both cut points, but only one (Figure 10.18a), then determine the location of the ear necessary as follows.

This situation means that in the rectangle of the cut, we know the coordinates of one of the four vertices through which the cut line passes.

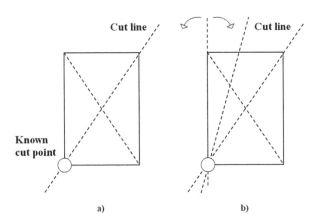

a)   b)

FIGURE 10.18
Determining the location of the cut line by the coordinates of one vertex cut rectangle.

We do not know which coordinates of exactly four vertices of the cut rectangle were found. It is not known the direction in which the face of the person is directed from the sides. Therefore it is necessary to apply a search from a certain number of options. To do this, a vertical line is drawn through the cut point, the coordinates of which are known. We will consider it as a cut line. We check the truth of its direction using the method of division into separate objects, which is described earlier. Then rotate this line relative to the known cut point at a certain angle relative to the vertical (Figure 10.18b). We use the method of division into separate objects. Repeat this order for all possible directions from 0 to $\frac{\pi}{2}$. We will consider the vertical direction (up) to be zero. We perform similar actions for directions from 0 to $-\frac{\pi}{2}$. The step of turning the cut line is chosen in accordance with the technical features of the system. The accuracy of determining the parameters of the image must meet the requirements of the tasks assigned to it.

All areas for which the conditions of the method of separation into individual objects are met should be considered potentially true areas of the location of the image of the ear and used in the process of further biometric identification. It is not possible to determine the areas of the ear location for which no coordinates of any cut point have been found. You should change the gradation of brightness in the process of obtaining the area of the edge of the ear or its shadow, or use some other methods to determine the desired area.

10.6 The Actual Application of the Method of Determining the Location of the Image of the Ear

The earlier-described method of determining the location of the ear image in the process of biometric identification can be significantly complicated due to the presence of a large number of noise inclusions. The wide variety of shapes that create a large number of wrong angles during the inversion and contour selection process will encourage you to search for the location of the ear among a large number of options. All available potential cut-off points should be divided into pairs, and each of them should be considered as a basis for determining the desired area. Also, in the presence of obstacles, each of the defined points of the section can be considered as one of the vertices of the rectangle of the section, in the case when only its coordinates are defined.

The method of finding the area of the image is based on determining the cut points on the image of the head in the profile. By methods of PST, these points are determined using wrong angles. Their search is possible in the presence of sharp inner corners of the figures. The acute internal angle of the inverted image can be obtained by inverting the original image, which has sharp angles. In PST, noise is considered to be individual pixels or objects with a small area. The sharp corners of the original image can be perceived as noise. In the process of removing noise by removing the contour points, such elements can be distorted. This will negatively affect the search for the wrong angles of the inverted image. Therefore, on real noise control images, the contour point removal method can only be applied to the inverse image. Other methods must be used for the initial image in the noise control process.

The second difficulty in applying this method to determine the location of the ear image is to select the brightness values of the image areas to select the area of the edge of the ear or the shadow around it. Subsequently, the same problem arises in the process of segmenting the location of the image of the ear for identification. The location of the identification devices and the environment create the preconditions for different lighting of the areas of the person's head that are analyzed. To deviate from the absolute values when choosing the brightness limits, they must be calculated relative to certain values of the elements of the current image.

When searching for a certain basic value of the brightness of the elements of the original image, keep in mind that the brightness of the areas of the head that are covered with hair, usually less than the brightness of the skin. The number of elements of the hairstyle in the image of the head in profile is approximately half of the area of the entire image. To determine the basic value of the brightness of the image, it must be divided into individual elements. The brightness of these elements must be sorted in ascending order and the average of the second half of the sequence calculated. The resulting brightness value can be taken as a unit and select the brightness value of the boundaries of the areas of the face image in accordance with this value.

Examples of finding the location of the ear image are shown in Figure 10.19.

For both examples (right ear Figure 10.19a and left ear Figure 10.19b), only one pair of points meets the criteria of this method among the found potential cut-off points when dividing the image into separate objects.

10.7 The Procedure for Comparing the Elements of the Ear Image with the Reference Data

You can change the base brightness value before the direct identification process. Because multiple cutouts can be found during the analysis, this can lead to the selection of multiple areas of the location of the ear image. However, if this area is selected, then a certain cut line is selected. Then for each case we choose a new value of the base brightness. This value can be the average

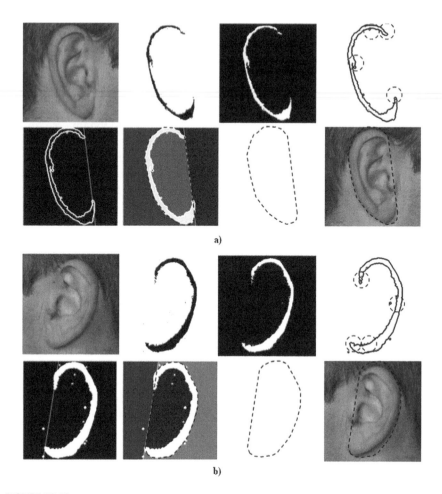

FIGURE 10.19
Examples of selection of the area of the ear elements.

value of a certain number of image elements from the section point (it must exist) in the direction along the section line. If the incision point exists, then it and adjacent areas are more likely to have a brightness close to the skin elements. Averaging is necessary due to the fact that in the areas of the selected elements there may be certain noise inclusions.

The image of the ear in a predetermined area is divided into several images. Separation occurs according to the specified brightness ranges (Figure 10.20a).

Reference images should be stored as multiple reference surfaces that correspond to relative brightness ranges. The number of these ranges for the standards and real images must be the same. Assuming that the image of the

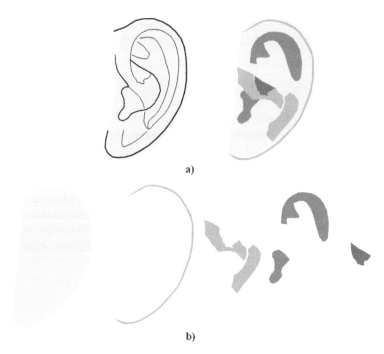

a)

b)

FIGURE 10.20
Divide the ear image into individual objects according to the brightness gradation.

ear is always equally oriented, then from the reference surface for verification in the process of identification may be sufficient indicators for orthogonal directions.

Each image created by elements within the same brightness range is treated as a separate object (Figure 10.20b). The ratios (rd_{Ai} and rd_{Bi}) of the dimensions of their bounding rectangles (A_{0i} and B_{0i}) to the dimensions of the bounding rectangle of the ear image area (A_{00} and B_{00}). are calculated. Sets of functions of the area of section for all images are formed.

$$\begin{cases} rd_{Ai} = \dfrac{A_{0i}}{A_{00}} \\ rd_{Bi} = \dfrac{B_{0i}}{B_{00}} \end{cases} \tag{10.5}$$

To image the ear for two orthogonal directions for all brightness ranges, sequences of size relations of the bounding rectangles are formed.

Areas of the ear image that fall into each of the brightness ranges are non-convex images of various shapes. Therefore, it is difficult to calculate the parameters of their bounding rectangles. These settings must be detected after dividing the ear image into brightness ranges. Reference parameters of size ratios ($rd_{etalonAi}$ and $rd_{etalonBi}$) should be stored in reference databases as additional parameters of standards.

The process of identification by the shape of the ear, as well as the processes of image recognition by PST (Bilan & Yuzhakov, 2018), should be divided into two stages (fast and detailed). This will speed up the identification process.

In the rapid identification step, the sequences of the calculated size ratios are compared with the corresponding sequences of the selected reference for the selected direction. If for all elements of sequence, there were a coincidence with the set accuracy, we pass to a detailed stage of comparison. At the detailed stage, the scalable functions of the intersection area of the standard and real images are compared. The comparison takes place with a given accuracy. If at any stage there is a discrepancy of values, we pass to check data of another standard. If all functions values for all ranges match, the identification process will be considered successful. Object identified.

11

Methods of Biometric Identification Based on Gait in the Smart City

11.1 The Main Reasons for Using Human Gait for Biometric Identification in a Smart City

As previously written, a person's gait is one of the important biometric characteristics that are used for biometric identification within a smart city. Analysis of a person's gait can be carried out at large distances and at various locations of the video sensors in relation to the human body while moving around the city.

Gait analysis is one of the dynamic methods of biometric identification. The application of methods of parallel shift technology to ensure the implementation of this process involves solving several subtasks. The first is to select an object from the background and/or among a group of similar objects for each frame of the video stream. The second subtask is to detect the motion vector of the object of study in the visual field and determine the conditional horizon for the image. The third task is to determine the period of a person's gait. The fourth subtask is to select the identification area from the object image. The fifth step of the process is to compare the data with the reference information.

Selecting an object from the background is the most difficult task to perform in the image processing process. In the case of using dynamic methods of biometric identification, its solution is complicated by the fact that similar actions for the same object must be performed for all frames of the video stream. The process of selecting a specific image from the background can be fully or partially interactive or automatic. In the case of interactive object selection, the operator must specify the criteria for its selection. The partial participation of an external observer in this process assumes that the selection rules are set in the initial section of the video stream, and the subsequent search for the object is done automatically based on the specified criteria. Full interactivity involves the constant participation of a person in the selection of a person in need of identification, throughout the video stream.

Full interactivity is more suitable for use in tracking systems rather than biometric identification. Interactive methods of selecting an object from the background are similar to the processes that take place in anti-tank systems. Partial interactivity is similar to "shot and forget" systems. Full interactivity is similar to "targeting" a target in laser-guided systems.

Automatically selecting an object from the background involves the use of sophisticated artificial intelligence techniques. These can be certain syntactic or probabilistic methods (Tou & Gonzalez, 1974). Syntactic methods involve the localization of the object of identification on certain signs of interaction with other objects (persons) of the visual scene or the structural features of oneself. Probabilistic methods are based on the assumption that some elements of the scene with a certain probability belong to the desired image. In both cases, the basis is the recognition of some elements of the image, the location of which is easier to locate (for example, people's faces), or details of other objects in the scene.

In object access systems, gait is not the main biometric parameter. However, the gait is effectively used in various systems for analyzing the conditions of various territories, which require close attention. Gait refers to dynamic biometric characteristics and it takes a certain amount of time to identify a person, which is sufficient for accurate identification. A person's gait is an irreplaceable biometric characteristic in conditions when it is impossible to fix other biometric characteristics, which is quite common in real life. There are many identification problems in which only gait data is available. Such tasks can arise in conditions of darkness, high dustiness, poor technical characteristics of video cameras, etc. When driving robots are used in an urban environment, robots must often select the right people, and for this, gait is often an indispensable characteristic.

As a result of the analysis of a person's gait over a certain period of time, an ordered series of quantitative values is formed that change at certain discrete points in time. The dynamics of changes in quantitative values is individual for each person.

Gait is one of the least researched parameters today (Kazantseva, 2013; Ghaeminia, Shokouhi, & Amirkhani, 2021; Bilan, Motornyuk, Yuzhakov, & Halushko, 2021e). It is fixed using special mechanical sensors or a video camera. Based on mechanical sensors, gait cannot be analyzed with just one sensor. To create and describe a complete image of a person's gait, it is necessary to use several sensors, which together describe various temporal, geometric, and power characteristics. For example, an accelerometer measures acceleration at specified time intervals, a pressure gauge measures the force of pressure on a surface. The bends of the legs and arms, the distance between the touches of the feet to the surface, the time between these touches, etc. are also measured. The more various non-uniform biometric characteristics are involved, the more information is involved in describing the gait, and the higher the identification accuracy.

However, the use of mechanical sensors requires additional means of data transmission, as well as high accuracy of signal fixation by each sensor. Mechanical sensors must first be attached to a person and use a special surface, which brings significant inconvenience to the person being identified. All the same, the multimodality of this approach gives higher accuracy.

A rather promising approach is the use of a video camera, since this makes it possible to analyze a person's gait (possibly less accurately) at significant distances. However, this method depends on the location of the camcorder and the use of large amounts of data.

The problems arising from the use of these methods sharply narrow the scope of the use of gait for human identification today. However, in world practice, there is a tendency to increase interest in the study of human gait, since in many problems this parameter is irreplaceable.

This paper describes the methods of biometric identification of a person based on the dynamics of his gait by analyzing the video sequence using theoretical positions based on the parallel shift technology, Radon transformation, and cellular automata with various forms of coverage.

The technology of parallel shift in image processing involves the analysis of the shape of objects. To do this, the image data must already be selected from the visual scene. Therefore, to consider the next stages of biometric identification along the way using the methods of TPN, we will consider the video series as a sequence of selected from the image of the silhouettes of one object. In this case, the process of selecting an object from the background is quite simple to organize. If the receiving device is stationary, then a permanent image without foreign objects can be considered as a background. When a specific object appears in the camcorder's field of view, elements of the new image set that match the background image elements are deleted. Everything left in the image is binary. This is the silhouette of the object of identification.

In modern research, the most popular are methods based on the analysis of a recorded video sequence of a person's movement, as well as using various sensors fixed on a person, a surface, or next to a person. This organization creates a multimodal biometric system.

Using video cameras, the biometric system becomes device independent.

At the same time, for greater reliability, several video cameras are used, located at different angles in relation to a person. For each video camera, various methods of data conversion and formation of characteristic features are developed.

As mentioned earlier, the gait changes for the same person if the properties and structure of the surface on which the person walks (asphalt, grass, sand, etc.) is changed. The geometric shape of a person is also changing for different locations of video cameras. However, in the process of walking, a person sooner or later enters a stable walking mode, inherent only to him.

When using video cameras, the main set of analyzed data is a video sequence of frames, each of which is an image of a given format. Using the generated video sequence, the period of repetition of human movements is determined. It is during this period that the main control states are recorded by which a person can be identified.

The main task of the method using a video camera is to improve the accuracy of identification regardless of the quality of the video data presentation.

11.2 Description of Gait Identification Method

The method is based on the analysis of a video sequence generated by a video camera. The video sequence reproduces the dynamics of a moving person. Cameras are positioned at different angles. It is enough to record the gait of the same person from three video cameras located at different angles (for example, from the side, top, and front) in order to accurately identify the walking person. In addition, it is assumed that the geometric shape of a person is known for all three angles. Also, to eliminate unnecessary information data, it is better to use binary video sequences. In such video sequences in each video frame, a walking person is represented by black (white) pixels, and the background, respectively, is represented by white (black) pixels. Such images can be obtained using conventional infrared (thermal) and other video cameras that produce two-color images. Figure 11.1 shows examples of video sequences of one person walking, recorded at angles 0°, 90°, and 36°. To obtain such sequences, we used the CASIA Gait databases, which are freely available (Zheng, Zhang, Huang, He, & Tan, 2011).

The well-known method for analyzing the video sequence of a moving person consists of the following actions (Bilan, Motornyuk, Yuzhakov, & Halushko, 2021e).

1. Binarization of all frames of a video sequence.

2. Selecting a moving person in each frame using a circumscribing rectangle.

FIGURE 11.1
Examples of video sequences of a walking person formed by infrared video cameras at angles 0°, 90°, and 36°.

3. Selection of a moving person in each frame of a video sequence, which have changed their color to the opposite in comparison with the previous frame.

4. Analysis of the area occupied by the selected pixels in each frame using the parallel shift technology and based on the analysis obtained, quantitative data are formed for each frame of the video sequence.

5. Data processing and formation of a quantitative image of a person's gait.

6. Comparison with reference values stored in the reference database.

7. An identification decision is made based on the confidence intervals used.

The first step of the method converts the video sequence to a binary form. This is done using special video cameras. The second step is carried out by well-known methods. These methods are based on determining the horizontal and vertical extreme pixels of a moving object. Through these pixels, straight lines (rows and columns of the image matrix) are drawn, the intersection of which outlines a rectangle, inside which the moving object is located. In each frame of the video sequence, such a circumscribing rectangle can be of different sizes.

In the third step, pixels are selected that have changed their color compared to the same pixels in the previous frame. They are highlighted to get the outline of a moving object and to reduce the number of pixels used for processing. The selected pixels are used to quantify the parallel shift technology. Based on this technology, area intersection functions are formed for each video frame. This is very convenient since there are no standard descriptions of the resulting shapes that form the selected pixels. The resulting FAIs are compared to obtain an overall relationship. This dependence, by its geometric shape, determines the dynamics of human movement.

On the obtained FAI curves, a section of the curve is selected that corresponds to the period of repetition of human movements. The length of this period is determined for each person. This period length is one of the main characteristics, according to the quantitative value of which the search for the standard in the standard database begins. An example of FAI formation for one frame of a video sequence is shown in Figure 11.2.

The repetition period was determined by calculating the average values for all FAIs that are obtained in the image of each video frame. On the basis of the obtained average values of the intersection area from the number of frames of the video sequence, a curve was built, the shape of which characterizes the gait of a person. The dependence of the average area of intersection of areas (the direction of the shift of the copy corresponds to $0°$) on the number of frames of the video sequence is shown in Figure 11.3.

On the graph, extremes indicate the greatest acceleration a person makes during one gait period. In these video frames, the human body has approximately the same shape. Neighboring video frames for the considered extreme

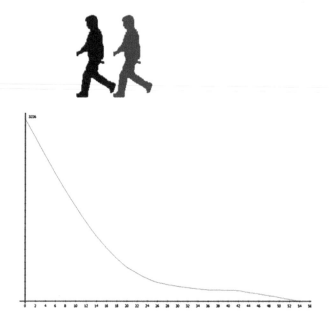

An example of forming an FAI for a single frame of a video sequence when the camcorder is
fixed at an angle 90°.

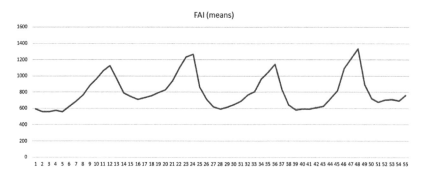

An example of the dependence of the average FAI value on the number of frames of a video
sequence for the shift direction 0°.

video frames have the same shapes, etc. Thus, for video frames located at a
distance of the repetition period, they have approximately the same shapes.

If the repeat period coincides with one of the reference values, then the
analysis and comparison of the average values within a certain period are
carried out. All comparisons are made taking into account the selected confidence intervals, which are determined experimentally for each location of

the video camera. The change in the size of the area determines the speed of approach or removal of a person from the video camera.

To find the repetition period, the extreme (maximum or minimum values) values in the obtained FAI sequence are searched for. The resulting number sequences are subtracted from the reference number sequences. If the received number G_{mean} does not go beyond the confidence interval $G_{mean} \in [-G;G]$, then for each such video frame 1 is formed, otherwise the value 0 is formed. Identification is carried out by counting the number of units formed as a result of comparisons for each video frame. Also, a threshold value for the number of units is set, which gives the identification of "our" or "alien". If the number of units exceeds the threshold value, then the person is identified as "our own", otherwise the person is identified as an "impostor". This identification is used for gaits with equal durations of repetition of movements in gait.

In the process of selecting cells, each pixel cannot always receive the same color codes, even if this pixel did not characterize movement and no changes occurred in its code. There are codes that do not visually display changes for human vision. But the program fixes them as pixels that have changed properties and selects them as participating in the movement.

To eliminate the reaction to visually close codes, a sensitivity threshold is introduced. Exceeding this threshold indicates that this pixel is involved in displaying motion and, accordingly, it is selected. For different sensitivity thresholds, a different number of pixels are allocated. Taking this circumstance into account, it is possible to realize the automatic setting of the required threshold. The system counts the number of pixels with the same codes, especially those pixels that encode a small amount of brightness are counted. Next, the number of pixels combined into luminance groups can be compared or the average value of the codes can be calculated and, based on such an analysis, the threshold value either increases or decreases. Also, the threshold value can be changed based on the analysis of the percentage of the number of pixels that captured the motion to the total number of pixels in one video frame. This takes into account the fact that the moving object is represented in advance by the set maximum size.

The sensitivity threshold can be changed if the percentage of the spread of the pixels that recorded the movement is determined.

The described method has the disadvantage that it takes a lot of time to process one video stream, since the parallel shift technology is used for each video frame. That is, on each video frame, there are shifts of copies of images that have different geometric shapes. Another disadvantage is that large amounts of data are generated for each person. These volumes are also time consuming to process.

The method described in the work allows to reduce the time spent on identification (Yuzhakov & Bilan, 2019). According to this technique, a parallel shift technology is used, which is realized by the natural shift of the image itself. The technique consists in the fact that all frames are sequentially superimposed on each other and the area of intersection of human images in two frames is calculated.

Frame overlays can be done for different options. The first option is that all frames are superimposed on the first video frame and the area of intersection of the human image is calculated on two images of the combined video frames. However, with a low frequency of video sequence formation, this technique may be unacceptable, as well as if the person is moving fast enough. Therefore, a different approach is used.

Another approach uses the analysis of images resulting from images of two adjacent video frames. For each video frame, the area of intersection of human images in all two adjacent video frames is calculated. If there are ten video frames in the sequence, then nine intersection areas will be calculated. An example of forming intersections for several frames of a video sequence is shown in Figure 11.4.

The more the video frames of a video sequence embedded in the period of repetition of human movements, the more accurate the identification result will be due to the larger amount of data in the function of intersection of the areas of adjacent video frames. Moreover, such FAI will be different in form from the form of the classic FAI. The intersection of areas with different shapes is described in Bilan & Yuzhakov (2018). Figure 11.5 shows an example FAI for two video sequences.

From the obtained FAI forms, it can be seen that for each person whose gait is analyzed, the FAI form has significant differences that clearly distinguish each person from another person. In this case, the frequency of forming video frames for all people should be the same. It is impossible to compare the gaits of several people, which are recorded by video cameras operating at different frequencies. The reference database should also be formed using video cameras with the same characteristics.

During the experimental studies, various confidence intervals were used and the optimal values for the database used were determined. For this database, a confidence interval has been defined that can correspond to values of 9, 10, and 11 pixels. The use of additional FAI, as well as the data of confidence intervals allowed to increase the percentage of correct identification for "our" to 78% and received less than 10% matches when identifying the "impostor".

FIGURE 11.4
Example of implementing intersections for multiple frames of a video sequence.

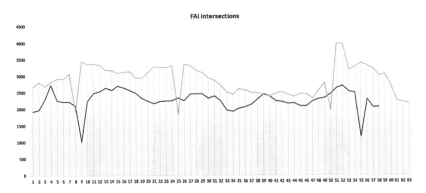

FIGURE 11.5
Example of FAI obtained from the intersection of images of neighboring video frames.

11.3 Search for the Motion Vector of the Object in the Visual Field and the Gait Period

Video streams are flat images that show a view of part of the three-dimensional space at a certain point in time from the location of the receiving device. The object of analysis in the process of biometric identification along the way is a sequence of images of the silhouette of the person. In the process of creating such a series, it is important to determine the relative position of the object of analysis and the receiving device (camera lens). Depending on the distance between them, the area of the silhouette image changes.

Due to the fact that the process of walking takes place in a horizontal plane (Figure 11.6), the reference information should be stored for certain points for each of the possible directions of movement of the identified person.

The location of the conditional horizon line depends on the height of the video camera lens above the plane in which the identified objects move, and the angle of the optical axis of the camera to one of the orthogonal directions. The location of the video camera in the process of biometric identification of a person on the move should be the same as the location of this camera during system training. The best location of the camera in height is such that the optical axis passes vertically in the middle of the body of a person of average height. This will allow you to perceive the silhouettes of objects of the maximum possible area for this direction of the optical axis (Figure 11.7).

You can also place the camcorder in places where the potential trajectory of potential identification objects is limited. However, the observation area must be of sufficient length to ensure that sufficient information is obtained to carry out the identification process.

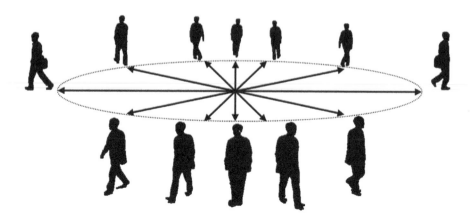

FIGURE 11.6
Possible directions of movement of the person.

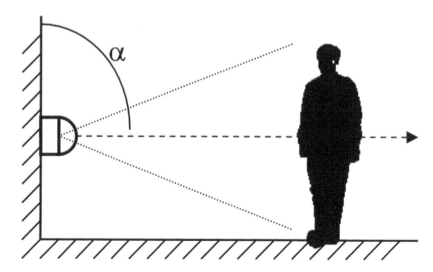

FIGURE 11.7
Location of the video camera for biometric identification by gait.

The motion vector ($\overline{\varphi}$) is formed by the horizontal displacement vector ($\overline{x}$) and the distance vector ($\overline{y}$).

$$\left|\overline{\varphi}\right|^2 = \left|\overline{x}\right|^2 + \left|\overline{y}\right|^2 \tag{11.1}$$

The calculation of the parameters of the horizontal displacement vector can be performed in two ways. The first method is described in previous works describing the methods of parallel shift technology in the sections (Belan & Yuzhakov, 2013b, 2018), which relate to the study of motion parameters. This method is based on the use of the Doppler effect. When organizing the cyclic motion of a copy of an image at a constant speed, the motion of the object in the visual field will affect the overall cyclic function of the intersection. If the object moves in the same direction with the direction of cyclic shift of its copy, the distance between the maximum values of the sections of the FAI, which corresponds to the intersection of the image of the silhouette and its copy, will increase. If the direction of movement of the object is opposite to the direction of cyclic movement of its copy, then this distance will decrease. If the width of the receptor field in the direction of the copy offset (*lenth$_x$*) and the distance between adjacent maxima of the total cyclic function of the intersection area (T_{xi}) are known, then the displacement modulus ($|T_{xi} - lenth_x|$) will be equal to the horizontal displacement vector ($|\bar{x}|$), and the sign of this difference object.

Information about the speed of cyclic shift of the image copy (v_{sh}) will allow to calculate the speed of the object (v_{obj}). The rate at which one frame is formed in the camcorder (τ) must be such that at least one new frame is generated while the copy of the full-shift image (*lenth$_x$*), image is traversed. If there are more such frames, the current location of the object image is fixed on the last of them. At a constant speed of the object $T_{xi} = const$.

Since $v_{sh} >> v_{obj}$, the distance T_{xi} can be taken not between adjacent maxima of the cyclic FAI, but between the first (existing at the initial time) and some j-th maxima. Then the formula of the modulus of the horizontal motion vector is slightly modified.

$$|\bar{x}| = |T_{xi} - (j-1) \cdot lenth_x| \qquad (11.2)$$

The value of the time of displacement of the object in space (t_φ) will be equal to the value of the time of displacement $|\bar{x}|$ by the horizontal distance (t_x).

You can search for horizontal offset vector parameters in another way. First, a certain k-th frame of the video is selected. The j-th frame of the video sequence is then selected so that the image of the object of study in this frame does not intersect with its image in the k-th frame. Images of silhouettes from both frames are combined. The image from the k-th frame is called A object, and the image from the j-th frame is called B object. There is a certain scene, the distance between the objects of which corresponds to the relative 1 or 2 positions (Figure 8.1). The distance between objects ($Dist_{AB}$) by formula (8.18) is calculated. This distance is equal to the value of the horizontal offset distance ($|\bar{x}|$). Note that the distance from $Dist_{AB}$ is determined from the right boundary of the object on the left to the left boundary of the object on the right. To more accurately determine this distance, it may be necessary to add the sum of half the width values of their bounding rectangles for the vertical shear direction. This need arises when the calculated distance between objects is comparable to their size.

The relative position of objects A and B in the receptor field will indicate the direction of displacement of the object of study. For these objects, the distance from the edge of the receptor field can be determined using a cyclic shift. Let's assume that the copy of objects moves to the right. Then, if the distance from the right boundary of object A (k-th frame) to the right edge of the receptor field is greater than the distance of the right boundary of object B (j-th frame) to the right edge of the receptor field, the silhouette of the object moves right. Otherwise it moves horizontally to the left. The time to move the object in space will be as follows.

$$t_\varphi = t_x = (j - k) \cdot \tau \tag{11.3}$$

Each of the approaches to calculating the parameters of the vector of horizontal motion has its disadvantages and advantages. The first is easier to implement, but because the parameters are analyzed in real time, the mutual movement of the copy of the image and the silhouette itself can slightly distort the data. The second approach is more difficult to implement, but allows you to analyze a static scene. These distances are a projection of real values on the region of the receptor field. Actual values must be obtained by calibrating the camcorder settings.

It is better to calculate the parameters of the motion vector taking into account such a parameter as the "gait period". Gait period (T_g) is the distance between adjacent identical phases of a person's movement (Figure 11.8a). The period is two steps of walking person. The main hypothesis, which is the basis for biometric identification of a person by analyzing his gait, is that gait indicators in certain areas of movement of the object are repeated. The

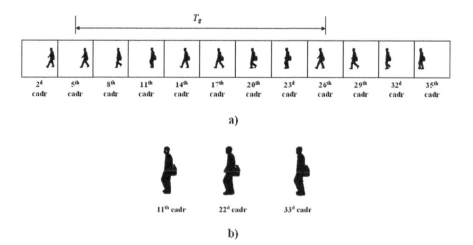

a)

b)

FIGURE 11.8
Sequence of frames of horizontal movement of the person.

image of the silhouette in different frames of the video stream at points of the same phase of movement, taking into account the scaling, is the same. Thus, in the calculations, the mentioned k-th and j-th frames should be chosen for the points that correspond to the distance of the object displacement in the horizontal direction on the projection of the period of the stroke for this direction (T_{gx}). Also among similar pairs of images, it is necessary to choose those where, taking into account scaling at formation of functions of the area of intersection for the vertical direction, the parameter of the maximum shift (Y_{max}) is the maximum. Another useful property is that this parameter at these points of the trajectory corresponds to human height. Let's call them "nodes of the trajectory". For the given example, such are frames 11, 22, 33 (Figure 11.8b). The gait period in this case approximately corresponds to the time of formation of 22 frames.

Here you should be careful. The parameters of the motion vectors ($|\bar{x}|$ and $|\bar{y}|$) are calculated for the directions relative to the plane of the trajectory, and the parameters of the functions of the intersection area are calculated relative to the orthogonal directions of movement of the image copy in the corresponding frame.

The modulus of the displacement vector of the face must be equal to the gait period.

$$|\bar{\varphi}| = T_g \qquad (11.4)$$

The modules of the orthogonal shift vectors are equal to the projections of the gait period on the corresponding directions of the plane of the motion trajectory.

$$|\bar{x}| = T_{gx} \qquad (11.5)$$

$$|\bar{y}| = T_{gy} \qquad (11.6)$$

If we take into account the fact that the average length of a person's step when walking is approximately half of his height, the gait period (T_g), which corresponds to the distance of two steps, is proportional to the maximum among the scaled maximum shifts (Y_{max}). However, this indicator is influenced not only by the features of a person's gait, but also the speed at which he moves.

The parameters of the distance vector $(\bar{y})$ can be calculated by analyzing the area of the silhouette image at different points in time.

First you need to determine the direction of movement. If the area of the silhouette decreases with time, then the object is moving away, otherwise the object is approaching. Area verification should be performed for several pairs of images separated in time by the gait period to reduce the effect of possible silhouette distortions.

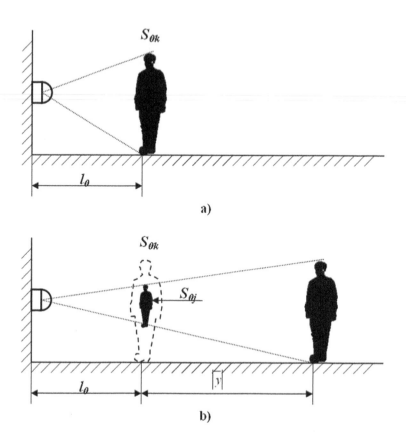

FIGURE 11.9
The location of objects relative to the camcorder at different times.

If there is an image from the k-th frame of the video sequence (Figure 11.9a), for which the distance to the object is l_0, and an image from the j-th frame (Figure 11.9b), for which this distance changes by $|\bar{y}|$, then, due to properties of proportionality, the equation will be as follows.

$$\sqrt{\frac{S_{0k}}{S_{0j}}} = \frac{l_0 + |\bar{y}|}{l_0} \tag{11.7}$$

In this case, the object is deleted. If it approaches, the formula will look like this.

$$\sqrt{\frac{S_{0j}}{S_{0k}}} = \frac{l_0}{l_0 - |\bar{y}|} \tag{11.8}$$

If the distance from the subject to the camcorder does not change, then $S_{0j} = S_{0k}$.

In the case of stationary technical means, the determination of the parameters of the distance to the object in the initial frame (l_0) can be performed by marking the surface elements or fixing the distance to certain background elements, which may be equal to the distance to the object at certain intervals (for example, doors). If you break the points of the nodes of the trajectory into pairs so that the second element of the i-th pair will be the first element of the next $(i + 1)$-th pair, you can build a chain that will correspond to the trajectory of the object. Determining the distance (l_0) for the first element of any of these pairs will find the coordinates for all nodes of the trajectory in the plane of motion.

Searching for the parameters of the motion vector is necessary in order to determine the direction of the reference information (α), which will be used in the comparison processes.

$$tg(\alpha) = \frac{\overline{x}}{\overline{y}} \tag{11.9}$$

Reference data should be stored as sequences of reference image surfaces for certain directions from 0 to π. This refers to the directions on the plane of the trajectory (Figure 11.7). The stored information has the property of axial symmetry. Images for the directions α and $(\alpha + \pi)$ are mirror images. Therefore, reference directions do not need to be stored for directions from π to 2π.

11.4 Defining the Area of Analysis in the Images of the Silhouette of the Object

The fourth subtask of the process of biometric identification of a person by gait analysis is to determine the area of analysis in the images of the silhouette. If the reference information about the person is stored in the form of images, instead of reference surfaces, then comparison of parameters of the course is possible with application of any known method. The use of parallel shift technology has its own characteristics. With the selected method of interaction of the object and its copy (search of intersection of sets) for processing of flat images manipulations occur with such parameter as the area.

The initial information for gait analysis is a set of flat images (video stream frames). Features of change of the area of a silhouette in the process of walking of each individual arise owing to movement of elements of its torso, arms, and legs. The head of different people in the image of the silhouette has approximately the same shape in the form of an ellipse. As a result, the image of this part of the body carries almost no useful information in the process of analyzing the area of the silhouette. Moreover, due to the possible presence of different hairstyles or hats, the silhouette can be significantly distorted.

The shape of hats can have a strong influence on the result of calculations. For example, a hat like a cylinder. In addition, during the various phases of a person's life, the size of the head can range from $\frac{1}{4}$ to $\frac{1}{8}$ part of its height. Information on the shape of the head does not apply to the process of identifying a person along the way. It is necessary to segment the silhouette to exclude the image of the head from the frame.

For such segmentation, it is necessary to look for the smallest section of a silhouette on width. This will help find the area of the neck image. However, the width of the arms and some legs is usually less than the width of the neck. However, the mentioned elements of the silhouette are absent in its upper third. Search for the neck area in the image of the silhouette of the face can be performed by highlighting the contour. The outline of the image can be selected by shifting the copy of the image. This method has been described previously. It allows you to select a contour of any width (Figure 11.10).

To find the area of the image of the neck, you must consistently increase the size of the contour. The first step in selecting a path determines the number of objects it limits. The outline should be four-pointed or its initial width should be two pixels to ensure that the internal elements of the silhouette binary image are separated from the background image. The initial number of elements of the silhouette will be equal to one.

Each step of searching for areas of the neck image consists of two parts. The first part is to increase the specified width of the contour and search for contour points of the silhouette by this parameter. The second part is to determine the number of individual elements of the silhouette, into which this contour divides the original image. If the number of such elements has changed, we check the fact of the appearance of a new separate area in the upper third of the silhouette. The number of individual elements of the entire silhouette may decrease when, found in a certain step earlier, the elements of the image areas of the arms or legs are included in the list of contour points.

In the example shown in Figure 11.10, the contour of the image with a width of 1, 3, 5, and 7 pixels is sequentially selected. When the contour width is 7 pixels, the initial image of the silhouette is divided into two separate elements. One of them is in the upper third of the image. Hence, this is the location of the image of the head.

FIGURE 11.10
The location of objects relative to the camcorder at different times.

If the width of the contour at the step of separating its elements is equal to $h_{contour}$, then the width of the neck image area will be equal to $2h_{contour}$. The image of the head should be excluded from the image of the silhouette. It is not required for the further process of biometric identification by gait analysis. However, this area of the image is already localized, so it can be used in the face recognition process (Figure 11.11).

In order to exclude from the image of the silhouette found elements of the image of the head, you must perform the following steps.

1. The binary initial image of the silhouette is saved.
2. All elements that are not included in a separate area in the upper third of the silhouette image, from the original image is removed.
3. Image inversion.
4. Selection of the contour width $h_{contour}$.
5. Image inversion. The area of the image of the head is obtained.
6. From the original image the items that included in the area of the head image are removed.

Items 3÷5 of this algorithm are similar to the combination of actions "IDI" of the noise control process by removing the contour points. This combination of actions will increase the area around the selected elements in the

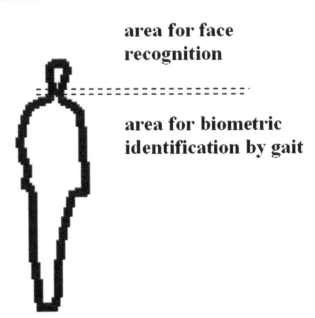

area for face recognition

area for biometric identification by gait

FIGURE 11.11
Dividing the silhouette image into two areas.

FIGURE 11.12
An example of the result of removing the image area of the head.

upper third of the image. The width of the contour selected in claim 4 of the algorithm may be greater than $h_{contour}$. However, increasing this parameter can significantly distort the part of the silhouette image that is selected for further analysis. An example of the result of extracting the image area of the head is shown in Figure 11.12.

The ratio of the height of the head area of the image to the height of the entire silhouette can be an indirect criterion for determining a person's age.

Therefore, to increase the accuracy of calculations in the implementation of the process of biometric identification by analyzing the gait for the formation of reference surfaces and to determine the parameters of the gait on the sequence of silhouettes of the object of analysis, it is necessary to remove elements.

11.5 Comparison of the Obtained Data with the Reference Information

The main feature of the process of comparing the actual obtained indicators of the functions of the area of intersection for images from each frame of the video stream with reference data is that the images of silhouettes of any person for any direction have small but almost identical contour parameter values (formula 6.7). This means that the study of such images by methods of parallel shift technology is possible, but the detection of differences between the FAI for different phases of movement requires the presence of high-precision instruments.

The second feature of this stage of analysis of forms of a sequence of silhouettes is that additional parameters, such as integral coefficients (formula 8.21), the ratio of parameters of bounding rectangles of each j-th frame to similar parameters of a certain initial k-th frame (system 10.5), and the coefficients of proportionality (formulas 8.24 and 8.25) will be close in values for the same directions of the image copy offset for different standards. Therefore, it is not advisable to use the fast stage of the comparison process, where additional indicators are compared. This process is limited to a detailed step, where the scaled functions of the area of intersection are compared.

At the learning stage, an image of the object's silhouette is selected in each frame of the video stream. The values of the motion vector and the gait period are determined. The head image area is then removed from each image. For each image, the functions of area of intersection are formed for all possible directions of its copy offset in the range from 0 to π. The parameters of each FAI are scaled.

To scale the values of the reference functions, a certain k-th frame of some range of the video stream with the smallest parameter of the initial area of the object is determined (S_{0k}). For each j-th frame from the selected range with the initial image area of the silhouette S_{0j} the scaling parameter is calculated (SC_{jk}).

$$SC_{jk} = \frac{S_{0j}}{S_{0k}} \qquad (11.10)$$

The values of the initial area will be the same for all reference images. In this case, the parameters S_0 act as one-dimensional quantities ($S_0 = FAI_\varphi(0)$). All parameters (area of definition and value) of functions of the area of section of each frame for each direction decrease in SC_{jk} times with preservation of proportions. Determined the number of full stroke periods (N_{Tg}), that fall within a given video stream range. Further manipulations are performed only with those with frames that are in this range from N_{Tg} periods of gait. Each stroke period is displayed on a certain sequence of m frames. For each i-th value ($FAI_\varphi(i)$) of each of the FAI for each direction of the image copy offset (φ) from each j-th frame the value of the averaging parameter is formed ($N_{j\varphi i}$). In the beginning, all these parameters are equal to the number of complete periods of the gait.

$$N_{j\varphi i} = N_{Tg} \qquad (11.11)$$

To obtain the average values of the functions of the area of untersection for each direction of the image copy offset of each frame, all the corresponding FAI values are summed and divided by the averaging parameter.

$$FAI_{javerage\phi}(i) = \frac{\sum_{a=0}^{N_{j\phi i}-1} FAI_{aj\phi}(i)}{N_{j\phi i}} \qquad (11.12)$$

There is a need to remove from the standard data that differ significantly from the average values. To do this, the values of the functions of the area of intersection for each direction of the image copy of each frame are compared, which reflects a certain phase of the stroke period ($FAI_{j\varphi}(i)$) with the corresponding average value $FAI_{javerage\varphi}(i)$. If they differ more than the specified value (δ), then most likely the image of this j-th frame is distorted. All functions of area of intersection for all directions of the image copy offset from this frame must be removed from the data that form the average values ($FAI_{javerage\varphi}(i)$). The corresponding averaging parameters ($N_{j\varphi i}$) are reduced by one. The corresponding average values are calculated by formula (11.12) taking into account the values of the new averaging parameters. This process occurs cyclically until all possible deviations become less than δ.

For example, a person in the process of movement tried to greet someone with a wave of the hand. She usually doesn't do this while walking. Images from frames that reflect this movement distort the reference data. Data from these frames must be removed from the calculations.

Thus, for a certain direction of movement of the person the sequence of reference surfaces which are created by scalable FAI from average values is formed ($FAI_{javerage\varphi}(i)$). This sequence corresponds to one period of gait for a given direction.

There is no need to average the video stream data obtained during the identification process. You need to determine the motion vector of the object. Then the k-th frame is determined from the video stream, which reflects the same phase of the gait period as the first frame of the reference sequence for the direction of movement, which corresponds to the found motion vector. Images from all frames of the video stream, starting with k, which reflect the gait period, are scaled. Scaling is based on the same principle as when creating a standard. The difference of this process is that the minimum value of the initial area is determined among the images of a given range of frames and parameters of any reference surface for the selected direction of movement (the values of their initial areas are the same). The data of the scaled functions of the area of intersection for all possible directions of the image copy offset of each frame of the selected range are compared with the corresponding data of the parameters of the reference surfaces. If all comparisons match the specified accuracy, the object is identified.

If there are other frames in the selected video stream range that show the travel period, similar sequences of scaling and comparison processes can be performed for them. This will increase the accuracy of identification.

The use of biometric identification methods based on PST requires high accuracy of calculations. They are more suitable for object classification processes. These methods may be less effective than other known methods of identification. However, their presence confirms the ability to solve various problems using parallel shift technology.

12

Biometric Data Processing of Human and Animal Colonies in a Smart City

12.1 Distributed Smart Sensor Network of Smart City

People (residents and guests of the city), as well as various animals (domestic and wild), live in the urban environment within the infrastructure. Both people and animals constantly move within the city itself, and can also go outside and contact other people and animals.

The principles of functioning of a smart city imply constant monitoring of the situation, since a smart environment should provide reliable protection of the city and its inhabitants from various threats. Also, constant monitoring of the state of the urban environment is necessary to maintain the necessary comfort of city residents, which is the basis of a smart city and it is constantly evolving toward increasing comfort.

A smart city environment must constantly respond to various events and requests of residents and visitors of the city. An important aspect of a smart urban environment is reliable protection, which consists in protecting human health and confidential information of a person, as well as protecting his property.

To protect human health and prevent situations that can lead to human injury, in a smart city, it is necessary to predict various events based on the current state of the smart urban environment. For example, it is very important in an urban environment to have a well-developed transport infrastructure to control the movement of vehicles and pedestrians, as well as to automate parking lots. It is also important to control the movement of people with a viral disease and for other tasks. Therefore, constant monitoring of the movement of people in a smart urban environment is one of the main tasks of the functioning of a smart city.

It has already been said that to solve this problem, the city must be equipped with a network of sensors located throughout the city and constituting an interacting multisensor system. If it is necessary to control the movement of people, then a significant part of the sensors should be aimed at fixing the biometric data of a person.

Biometric sensors help control the following tasks:

- the spread of viral diseases;
- access to objects of collective and private property;
- belonging to a given community, united by the selected common properties;
- protecting a person from unwanted meetings with other people.

The control of viral diseases is carried out as follows.

A large number of people live in the city, who have their own quantitative values for each biometric characteristic. If one of the residents or guests falls ill with a viral disease that is transmitted by airborne droplets, then in this case his biometric characteristics are recorded, which are recorded in the city digital network. A fixed sick person is prohibited from leaving the apartment and contacting other people. If this person has left the apartment in an urban environment, then biometric sensors should track his movement and warn him of meetings and contact with other healthy people. In this case, each biometric sensor that has detected a person with a viral disease must send a corresponding signal to the system. The system, which received such a signal, determines the location of such a sensor and transmits signals to neighboring signaling devices and mobile devices of other residents, who are notified of the movement of a sick person. Also, using messages, the city information system forms the desired route of movement of healthy people.

Access to objects of collective and private property is realized using various biometric sensors, which are located at the access entrances to objects. If this is an object of private sole property, then biometric authentication of the person who owns this object is carried out at the entrance. Authentication does not require large amounts of memory, which stores reference vectors of characteristic features of a biometric image of only one owner of an object. Human authentication algorithms are not complex and they yield a very high percentage of accurate authentication. Additional biometric characteristics are also used, which leads to almost 100% authentication. Typically, a fingerprint is used as the initial primary biometric characteristic, and dynamic characteristics are used as additional biometric characteristics. These additional biometric characteristics often include the dynamics of handwriting and keyboard handwriting, as well as other characteristics.

Biometric identification of a person is used to access shared facilities in theurban infrastructure. Biometric identification systems use a large memory to store reference vectors of characteristic features of each biometric image. Also, complex identification algorithms are implemented that classify biometric images. If only one biometric characteristic is used, then there is a certain percentage of false identification, which is much higher than the percentage of false authentication (for one user). Therefore, biometric identification systems also use additional biometric characteristics. These additional biometric characteristics can also be static and dynamic. At the same time, all additional characteristics should be recorded by non-complex biometric sensors. It is desirable that they belong to the same system. For example, biometric characteristics are entered using a keyboard, mouse, tablet, or by other means of entering information into the system.

Belonging to a given community, united according to the selected common properties, is carried out using biometric characteristics that are predetermined at the time of identifying or assigning properties to a person. A part of the memory of the general system is allocated, in which reference vectors of characteristic features are stored, which identify people with given properties. Each property is defined by a specific set of biometric characteristics. According to this set of biometric characteristics, a selected number of people are endowed with certain access or priority properties. These people may have general physical abilities in relation to other residents of the city.

Protection of a person from unwanted meetings with other people is carried out by determining the biometric image of an unwanted person. This image is distributed across the entire sensor network and transmitted to those people on mobile devices who should not meet with the identified person. Its location is transmitted.

In a smart city system, people can move by different means. Nowadays, the means of radio-technical tracking of movement of people and automatic vehicles are widely used (Klems, 2005; Symond, Ayoade, & Parry, 2009; Bhuptani & Moradpur, 2007). For this, special radio frequency tags are used on each tracked object. This is the main disadvantage, since the loss of radio frequency markers is possible or various physical influences can affect them, which lead to various kinds of distortion. RFID does not use biometric data, so it cannot be highly reliable.

The use of biometric characteristics divides the city's territory into sectors, which are analyzed by video cameras, or by a set of specially and orderly located biometric sensors in each sector. Each sector can be analyzed by one video camera, the viewing area of which is covered by this video camera. The number of sectors corresponds to so many cameras. Figure 12.1 shows a diagram of the division into sectors of the territory of a smart city.

The general visual picture of the urban area is formed on the basis of video data received from video cameras that analyze the territory of each sector of the city. The overall visual picture can only display the territory of one sector, several sectors, and all sectors at the same time. The movement of all objects in all sectors at the same time can be displayed. However, a situation may arise when it is necessary to identify several moving people who move in different sectors among many people walking. It is possible to do this using biometric sensors. These sensors identify specified people across all sectors in which they are located. An example of such a display is shown in Figure 12.2.

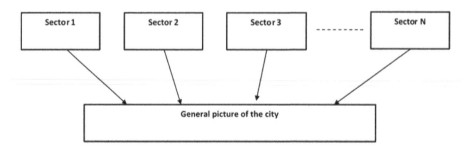

FIGURE 12.1
Dividing the city into sectors.

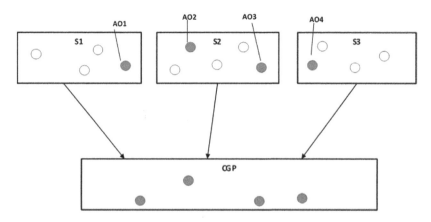

FIGURE 12.2
An example of the identification of four people recorded in three sectors of the city.

Shaded objects in each sector (Si), using video cameras, are transmitted to the general picture of the city (CGP), where they are displayed together and without those people who are not identified (objects not shaded in each sector). All shaded objects move in each sector, and their displacement (movement) is displayed in the general picture of the city.

12.2 A Method for Analyzing the Overall Urban Picture of the Behavior of People and Animals in a Smart City

With such an organization of the system for displaying the movement of people, there is the task of correctly representing and describing the visual scene on CGP. The image of moving people can be understood by humans. However, it is difficult to represent this in an automated smart city system.

In this work, such an image is described by a pulse train. Time is used – an impulse method of representing each geometric object, as well as an image containing several geometric objects (Bilan, Bilan, & Motornyuk, 2020).

For such a description, one geometric object (its geometric shape) is first described. The object is described based on the technologies of cellular automata with active cells using control transmission of an active signal (Bilan, Bilan, & Motornyuk, 2020).

This technology is as follows.

The object image is displayed on the cellular automaton environment with orthogonal coverage. The cells of the figure are displayed in black (black is encoded with a logical "1"), and the cells in the background are displayed in white (white is encoded with a logical "0"). With each time step, all image cells do not change their informational state. Only cells with an active state change their state. Each cell of such a cellular automata can have two states:

- basic informational state;
- additional active state.

The main information state is determined by the execution of the logical function of states (LSF). The LSF arguments are signals from information outputs of cells belonging to the neighborhood of the cell under consideration. Neighborhoods can be represented in different forms. The most popular forms of neighborhoods are the von Neman and Moore neighborhoods (Bilan, Bilan, & Motornyuk, 2020; Bilan, 2017).

An additional active state is determined by the execution of the logical transition function (LTF). The arguments of this function are signals from informational or from active outputs of cells belonging to the neighborhood. From the active exit of a cell that have an active state, an active signal is transmitted to that cell of the neighborhood, which was indicated by the LTF of the active cell.

In a CA with controlled transmission of an active signal, an active signal is transmitted only through those cells that are in a certain state and have neighboring cells with a defined state. For our case, such cells are black cells belonging to the contour of a geometric figure.

Figure 12.3 shows an example of the formation of a pulse sequence of one geometric figure.

FIGURE 12.3
An example of the formation of an impulse sequence of one geometric figure.

At the initial moment of time, the initial active cell is selected, which belongs to the background. With each time step, the number of active cells increases in accordance with the selected neighborhood. In Figure 12.3, the Moore neighborhood is used. When one of the adjacent background cells of the contour becomes active, all active cells of the background are zeroed (pass into a passive state). One of the neighborhood cells of the contour to the active cell of the background goes into an active state. From this moment on, an active signal begins to spread through the cells of the circuit. With each transition of an active signal to a neighboring cell of the contour, an impulse is formed. If the adjacent point of the contour lies on the same straight line with the two previous active cells, then a pulse of unit amplitude is formed. If the contour cell is located at the convex break point of the contour of the geometric figure (the convex vertex of the geometric figure), then the amplitude of the generated pulse increases. The smaller the angle at the vertex, the greater the amplitude of the positive pulse. If the vertex is concave, then a negative polarity pulse is generated. After the active signals pass all the cells of the contour, an initial pulse sequence is formed, which contains many single pulses (contains a unit amplitude).

Amplitude threshold processing is applied to such an initially generated pulse train. As a result of this, a pulse sequence is formed, containing as many pulses as there are vertices in a geometric figure. Moreover, the amplitudes of these pulses are equal to the number of unit remote pulses between two tops.

If there is more than one geometric object in the visual scene, then a more complex method is used. The method is as follows.

When the image is prepared and rendered into a binary form, the basic data mining operations are performed:

- selection of objects in the visual scene;
- outline description of each selected object;
- determination of the anchor point of each selected object;
- according to certain anchor points, the locations of each selected object are determined;
- ordering of all selected objects in the visual scene;
- description of the image and the formation of a descriptive pulse sequence.

12.3 Selection and Description of Objects in the Visual Scene

The selection of objects in the visual scene is carried out by propagating an active signal over the image field, projected on a CA with orthogonal coverage. The active signal, propagating through the cells belonging to the

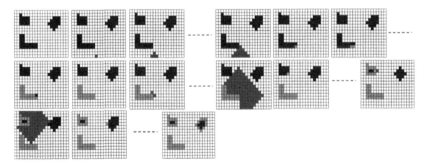

FIGURE 12.4
An example of selecting objects in a visual scene.

background, reaches single cells of the contour and sequentially transfers them into an active state (Figure 12.4). All cells belonging to the image of each local object in the scene have the same properties (all have an informational state of the logical "1"). The propagation of an active signal along the cells of the object's image contour integrates them into a single local geometric object with the same properties. How integration into one object is carried out is described in the work by Bilan (2019).

Such images take into account that the object has no discontinuities. This means that in each object it is possible to transmit a signal from one cell to any cell that belongs to this object. This suggests that each cell has at least one of its cells in the neighborhood. There are direct and indirect connections between all cells. Direct connections are determined by connections with the cells of the neighborhood, and indirect connections are determined by the fact that signals are transmitted through the cells located between two cells of the object. If a cell transmits an active signal to a distant cell of the same object, then all the cells through which the active signal passed will become linked into one object.

In this case, the cells of different objects cannot exchange signals, since they are separated by the background cells. In complex images (consisting of several objects), the sequence of selecting objects does not depend on the number of cells that form the object, depends on their location in relation to the first active cell.

Figure 12.4 shows that at first the first active cell of the background (from the bottom line of the CA) was selected and at the next time steps the active signal propagated through the cells of the background using the von Neumann neighborhood (highlighted in orange). At the fifth time step, the active signal reaches the nearest cell of the circuit, which is set to the active state (highlighted in green). At this time step, a pulse signal is generated at the general output of the CA. At the next 15 time steps, the active signal propagates only along the outermost cells of the first object (highlighted in green). An impulse is formed at each time step.

When the active signal reaches the first active cell of the outline of the first image (15th time step), at the next time step, the active signal is transmitted only to neighboring cells of the background (highlighted in orange). From this time step, the active signal begins to propagate through the background cells until it reaches the nearest cell of the contour of one of the not yet selected objects (24th time step in the general functioning of the CA from the beginning).

Starting from the 25th time step, the active signal begins to propagate only along the edge cells of another image object. A pulse sequence is formed, which described the second selected object. After the description of the second object, the third object is selected and the third pulse sequence is formed.

12.4 Time Impulse Description of the Visual Picture of a Smart City

There are also other options for selecting objects that implement bidirectional propagation of an active signal along the cells of the contour, as well as a method for removing the active cell of the contour when transmitting the active signal to the next cell of the contour. These methods allow to increase the speed of traversing the contour, but complicate the computational operations of the image description.

When forming a pulse sequence, different quantities and properties are used. This method takes into account the size (area) of the object and the distance between them. The distance is determined by the propagation time of the active signal along the background cells from object to object. In the method, at the initial moment of time, the distance to the first geometric object is determined. Each object is represented by three numbers. These numbers indicate the number of cells, distance, and location. Based on the data obtained, an impulse sequence of the entire visual picture is formed. Each such pulse sequence is also supplied in the form of a digital sequence of numbers that reflect the directions of transmission of the active signal through the image cells. An example of the generated common pulse train is shown in Figure 12.5.

A pulse sequence consists of several groups of pulses separated by long pulses that indicate the distance between objects. As you can see, the width of each such pulse differs from the others. Each group of pulses starts from one thin pulse, the amplitude of which indicates the position of this object in relation to the first object of the generated pulse sequence.

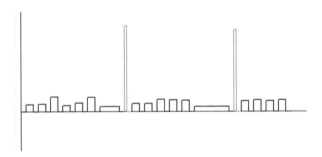

FIGURE 12.5
Example of a common pulse train after thresholding.

In each group of pulses, the number of pulses, their amplitudes, and polarities indicate the geometric shapes of each object. The form of the pulse sequence is also affected by the location of the first (largest in area) object in the visual scene, as well as the location of other objects.

In addition to the pulse sequence, the scene can be described using an ordered set of numbers (Figure 12.6).

The visual scene is represented by several groups of numbers, which have a strict arrangement. For such a strictly ordered sequence of numbers, it is possible, in automatic mode, to determine the following characteristics.

- the number of objects in the visual scene;
- the size (area) of objects;
- distances between objects;
- the location of objects in the visual scene.

In accordance with this method, software has been developed that allows you to describe the state of the location of moving objects on the territory of a smart city. The described method considers the state of the location of moving controlled objects at a fixed time. At the same time, if objects are moving, then at each subsequent moments of time, the CGP of the smart city changes. The general impulse sequences formed at the corresponding moments of time also change.

To simplify the algorithms for generating a pulse sequence, all moving objects are represented by describing polygons. For orthogonal coverage, such objects are most simply represented by circumscribing rectangles. In this case, groups of pulses in a pulse sequence will consist of four pulses of positive polarity. It is also possible to use other shapes describing polygons. For example, each individual controlled person can be represented by a separate polygon or a polygon is set for each sector.

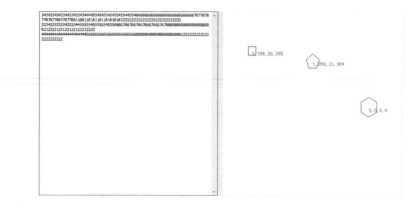

FIGURE 12.6
Example of a common pulse train after thresholding.

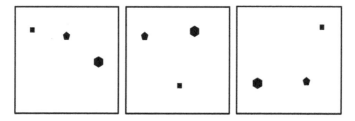

FIGURE 12.7
An example of changing the location of objects in the visual scene at different time steps.

Figure 12.7 shows an example of moving objects in the CGP at three fixed time steps. These steps do not constitute a separate fragment of adjacent video frames from the captured video sequence. Video frames recorded at different times are considered.

In this example, an object from the corresponding sector is represented by its describing polygon. Objects are positioned differently in each video frame. Accordingly, the forms of the pulse sequences change for each video frame (Figure 12.8).

Figure 12.8 shows different shapes. The automated system of the smart city automatically searches for differences between pulse sequences describing each video frame of the CGP. The obtained quantitative values of the differences allow the automated intelligent system of a smart city to make decisions for subsequent actions to improve comfort and increase the protection of people in a smart city.

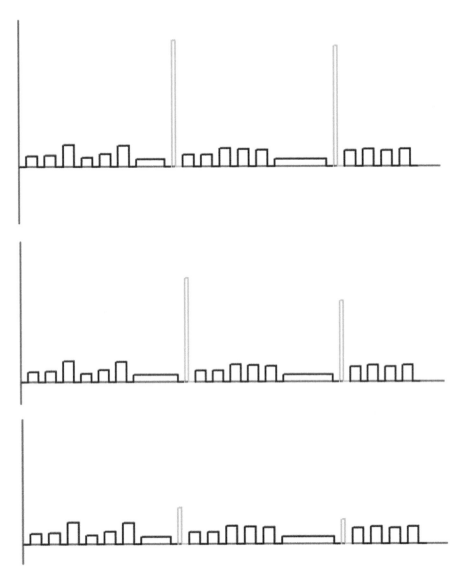

FIGURE 12.8
Forms of impulse sequences describing video scenes shown in Figure 12.7.

The developed software also makes it possible to classify visual scenes. For this, the system has a learning mode, in which reference vectors of visual scenes are formed, which are remembered in the memory of reference standards. Among the reference vectors, there are also standards that indicate an undesirable situation in the territory of a smart city. When undesirable situations appear on the CGP, the system signals this and generates a sequence of actions to eliminate it.

The described method has a drawback, which is the impossibility of constructing a description picture if the objects do not intersect. However, this does not apply to this system, since the city is divided into non-intersecting sectors, as well as people who move and do not collide with each other.

13

Behavioral Models of Human and Animal Colonies Based on the Technology of Cellular Automata with Active Cells

13.1 Theoretical Positions of Cellular Automata Technology with Active Cells

Cellular automata (CA) with active cells can be used to implement models and paradigms that describe the dynamics of a smart city functioning with great efficiency (Bilan, 2017; Bilan, Bilan, & Motornyuk, 2020). CA with active cells represent new paradigms among the whole variety of CA paradigms. The first detailed works in this direction were published in 2017 (Bilan, 2017) and more extensively published in 2020 (Bilan, Bilan, & Motornyuk, 2020). These works describe the theoretical foundations of CA technologies with active cells.

CA with active cells is a homogeneous cellular structure that belongs to the family of asynchronous CA (ACA). Each cell of such an ACA differs from all known classical CAs in that it has two states: main and additional. The main state determines its informational state (logical "0" or logical "1"). The main informational state is determined by the LSF, as in most binary CAs. The main informational state of each CA cell depends on the states of the cells in the vicinity of each cell. In such ACA, not all cells perform LSF at every time step. LSF is performed only when the cell is in an additional active state.

A cell can become active at the next time step if at the current time step one of its neighborhood cells is active. The active cell is established at the initial moment of the CA functioning. In this case, there may be several active cells. The number of active cells can increase or decrease.

The cell structure of such an ACA contains an information output and one or more active outputs. The information output displays the informational state of the cell at the current time step. Active output or outputs are intended for the formation and transmission of an active signal.

If a cell contains several active cells, then their number must be equal to the number of cells in the neighborhood. Each active cell output is connected to the active input of the corresponding neighborhood cell. In this mode, the active cell selects the neighborhood cell that will become active at the next time step. At the corresponding active output of the active cell, a single signal appears, incoming at the active input of the selected cell in the neighborhood of the active cell. At the next time step, the selected neighborhood cell becomes active. The active cell at the current time step selects the next active cell based on the analysis of information states of the cells in the neighborhood of the active cell. The active cell performs both LSF and LTF. Inactive cells remain in their previous state.

If an active cell contains one active output, then an active signal is transmitted to the active inputs of all cells in the neighborhood. Each cell in the vicinity of an active cell analyzes the state of cells in its vicinity and performs LTF. If, as a result of performing LTF, a single signal is formed, at its active input there is also a logical "1" signal, then at the next time step this cell goes into the active state and performs LSF.

Thus, at each time step, only active cells change their basic informational state. The rest (inactive) cells do not change their state. Also, at each time step, the previous active cell becomes passive, and one or several of its neighboring cells become active.

For the normal functioning of ACA with active cells, it is necessary to make the correct choice of the following components of the ACA model:

- local function of states;
- local function of transitions;
- the form of the neighborhood;
- initial state.

The choice of LSF affects the duration of the functioning of the ACA. There are LSF (for example, logical functions AND, OR), which gradually bring all cells of the ACA to one state, which stops the normal functioning of the ACA. However, a successful choice of only LSF will not solve all the problems in organizing an ACA. A big problem is the problem of a successful choice of LTF. When choosing a LTF, it is necessary to take into account the situation so that there are no repetitive cycles. An ideal LTF is a function that randomly chooses the next active cell in the neighborhood. For example, if after the transmission of an active signal to one of the neighborhood cells at the next time step, nothing has changed in the state of the previous active cell, then if the active signal returns to the previous active cell, the new active cell must change its state in order for the active signal to pass to another cell in the neighborhood.

If the LTF is selected in such a way that sooner or later local cycles arise, then it is necessary to provide for additional actions to change the information state of the neighborhood cells, which sooner or later forces the ACA to change the direction of transmission of the active signal. A constant change in the state of cells in the neighborhood of an active cell eliminates looping. The dynamic environment is always changing. The whole world is constantly changing.

Another important characteristic of an ACA with active cells is the shape of the neighborhood and the number of cells that make up the neighborhood. The more cells are in the neighborhood of an active cell, the less the likelihood of active signal transmission cycles in the ACA field. If the neighborhood cells are not the nearest neighbors, then the active signal can jump over several cells at once.

It also depends a lot on the initial state of the ACA and the initial location of one or more active cells. The more active cells are used in an ACA, the longer the ACA functions without completely looping all the cells in the neighborhood.

If an ACA contains several active cells, then there is a possibility of combining several active cells in one ACA cell. In this case, the number of active cells decreases by one cell if the LTF of the matched active cells are the same. Ultimately, only one active cell can remain in the process of functioning. To eliminate this situation, different ACA cells are used in active cells by increasing the number of memory elements encoding the corresponding active state. Thus, a cell can have different active states, which are determined by different LTF. The bit output of the cell increases. The number of active states of a cell is limited by the maximum possible amount of LTF that can be realized in one cell. The maximum possible amount of LTF is limited by the number of neighborhood cells that are analyzed by the active cell.

13.2 Interaction of Active Cells

The paper by Bilan, Bilan, and Motornyuk (2020) describes how active cells interact with the same and different active states. Figure 13.1 shows three modes of interaction of active cells.

Figure 13.1 shows three modes of interaction of active cells that have a Moore neighborhood (eight neighborhood cells). Three modes of interaction are shown:

1. Interaction mode of two active cells in one cell of ACA cells.
2. The mode of interaction of active cells involves the fact that at a certain step, two active cells become neighborhood, i.e. one of the active cells is a neighborhood cell for the other and vice versa.

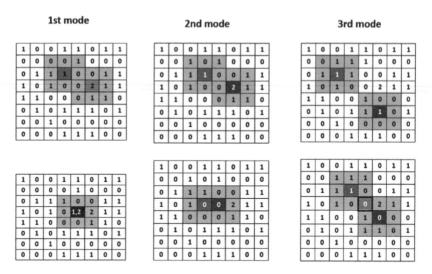

FIGURE 13.1
Modes of interaction of active cells.

3. A mode in which active cells interact with the cells of their neighborhoods (they have common cells belonging to the neighborhoods of both cells).

The first mode involves the complete combination of both cells in one cell. Such a complete combination is possible only if their active signal transmission is performed by different LTFs or the same LTFs, but with different function arguments. In this mode, both cells can become inactive at the same informational states of active cells and cells of the neighborhood at the previous time step. If the states of the neighborhood cells at the previous time step were different, then at the current time step, a new active cell with an active state is "born", which depends on the states of the cells of the neighborhood of two active cells at the previous time step. At the next time step, two active cells diverge according to their LTF, and a new active cell appears in the cell where these two active cells coincide. The new (third) active cell performs another LTF. One of the options for the formation of a new LTF is described in the book by Bilan, Bilan, and Motornyuk (2020). This paper describes coding options for new combinations of neighborhood cells for von Neumann neighborhoods and Moore neighborhoods. This approach assumes a large variety of combinations of different LFP depending on the states of the LFP. Figure 13.2 shows an example of forming a new active cell for the first variant of matches.

For each active cell, three neighborhood cells are highlighted in different colors, the codes of which indicate the direction of movement of the active

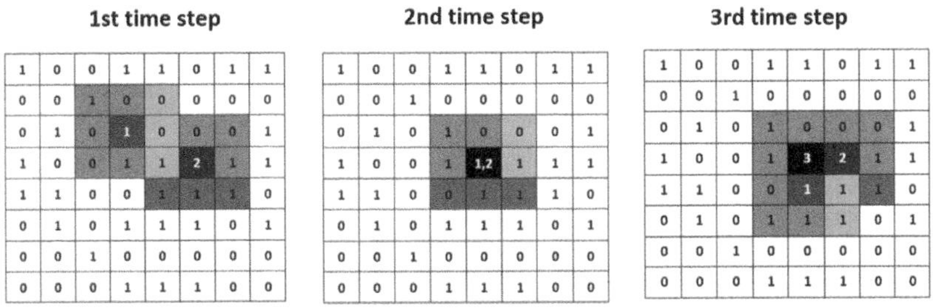

1st time step

1	0	0	1	1	0	1	1
0	0	1	0	0	0	0	0
0	1	0	1	0	0	0	1
1	0	0	1	2	1	1	1
1	1	0	0	1	1	1	0
0	1	0	1	1	1	0	1
0	0	1	0	0	0	0	0
0	0	0	1	1	1	0	0

2nd time step

1	0	0	1	1	0	1	1
0	0	1	0	0	0	0	0
0	1	0	1	0	0	0	1
1	0	0	1	1,2	1	1	1
1	1	0	0	1	1	1	0
0	1	0	1	1	1	0	1
0	0	1	0	0	0	0	0
0	0	0	1	1	1	0	0

3rd time step

1	0	0	1	1	0	1	1
0	0	1	0	0	0	0	0
0	1	0	1	0	0	0	1
1	0	0	1	3	2	1	1
1	1	0	0	1	1	1	0
0	1	0	1	1	1	0	1
0	0	1	0	0	0	0	0
0	0	0	1	1	1	0	0

FIGURE 13.2

An example of the interaction of two active cells and the formation of a new active cell for the first mode of coincidence of active cells.

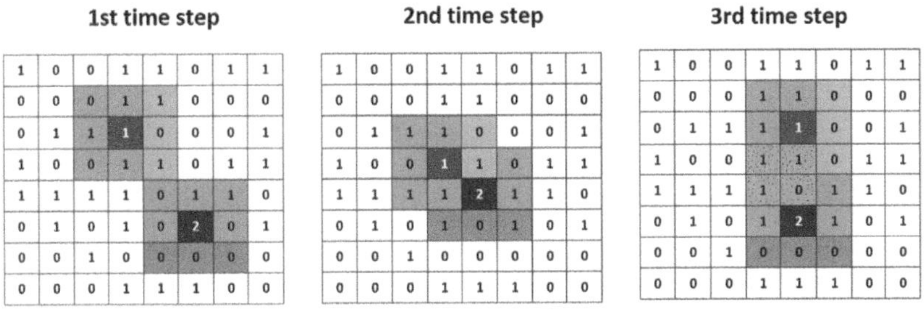

1st time step

1	0	0	1	1	0	1	1
0	0	0	1	1	0	0	0
0	1	1	1	0	0	0	1
1	0	0	1	1	0	1	1
1	1	1	1	0	1	1	0
0	1	0	1	0	2	0	1
0	0	1	0	0	0	0	0
0	0	0	1	1	1	0	0

2nd time step

1	0	0	1	1	0	1	1
0	0	0	1	1	0	0	0
0	1	1	1	0	0	0	1
1	0	0	1	1	0	1	1
1	1	1	1	2	1	1	0
0	1	0	1	0	1	0	1
0	0	1	0	0	0	0	0
0	0	0	1	1	1	0	0

3rd time step

1	0	0	1	1	0	1	1
0	0	0	1	1	0	0	0
0	1	1	1	1	0	0	1
1	0	0	1	1	0	1	1
1	1	1	1	0	1	1	0
0	1	0	1	2	1	0	1
0	0	1	0	0	0	0	0
0	0	0	1	1	1	0	0

FIGURE 13.3

An example of the interaction of two neighboring active cells and the formation of a new active cell for the second mode of interaction of active cells.

signal at the next time step. At the third time step, the matched two active cells transmit an active signal to the cells according to the codes of the cells of the neighborhood, and a new active cell (highlighted in blue) is formed in the cell where the two previous active cells coincide.

The second mode of interaction of two active cells implies the interaction of neighboring cells belonging to the neighborhood and does not imply a complete combination of two active cells. In this mode, the probability of meeting two active cells is greater than in the previous mode. Figure 13.3 shows an example of the formation of a new active cell in the second mode.

The example shows the interaction of two active cells and the formation of a new active cell. Shaded cells indicate that a new active cell may appear in one of these cells. The options for the appearance of new active cells can be different and are set in advance. Figure 13.3 shows an example where the initial two cells are active. At the third time step, they diverge and are not adjacent (the neighborhood does not belong to each other).

The disadvantage is the possibility of constant interaction of active cells, since cells at many time steps can be adjacent. Also, new active cells constantly interact with other active cells and form new active cells. In this case, it is necessary to introduce additional restrictions on the formation of new active cells. It is obvious that this mode can lead to an avalanche propagation of active cells and in a short number of time steps all ACA cells become active.

The third mode of interaction of active cells has an even greater probability of meeting and interaction with active cells. The avalanche effect of the propagation of active cells is also present in this mode and also leads to a rapid transition of all ACA cells into an active state. This mode may also be unacceptable for the normal functioning of the ACA. Since the avalanche effect in this mode is obvious, no graphic confirmation of this mode is provided.

At the same time, the well-known literary sources consider the modes of formation of new active cells (Bilan, Bilan, & Motornyuk, 2020), but little attention is paid to the modes in which the restriction of unwanted interactions of active cells is introduced. The second and third modes are the most suitable for implementing restrictions on unwanted meetings of active cells.

If it is necessary to use unwanted meetings of active cells, then the third regime is the best mode of their implementation, which limits the approach of other active cells to unwanted cells. This mode can be used to simulate the processes of combating viral diseases, as well as when simulating military operations and other processes. In a smart city environment, unwanted meetings can be prevented for people with a viral disease, wild animals, people with criminal intent, and other cases.

To avoid unwanted encounters, it is necessary that in the process of movement of active cells they receive signals from the cells of the neighborhood about the permissible directions of transmission of active signals. In this mode, the cell of the neighborhood generates a signal to prohibit receiving an active signal. Such a cell can generate a prohibition signal if it is a cell in the neighborhood of the cell that belongs to the neighborhood of the cell of an unwanted active cell. This situation is shown in Figure 13.4.

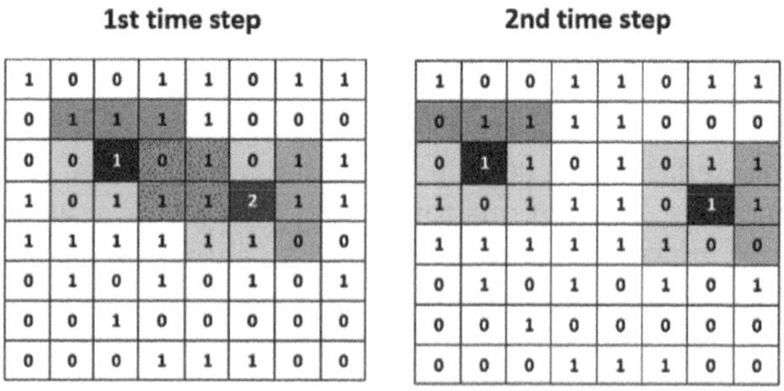

1st time step

1	0	0	1	1	0	1	1
0	1	1	1	1	0	0	0
0	0	1	0	1	0	1	1
1	0	1	1	1	2	1	1
1	1	1	1	1	1	0	0
0	1	0	1	0	1	0	1
0	0	1	0	0	0	0	0
0	0	0	1	1	1	0	0

2nd time step

1	0	0	1	1	0	1	1
0	1	1	1	1	0	0	0
0	1	1	0	1	0	1	1
1	0	1	1	1	0	1	1
1	1	1	1	1	1	0	0
0	1	0	1	0	1	0	1
0	0	1	0	0	0	0	0
0	0	0	1	1	1	0	0

FIGURE 13.4
An example of the possible proximity of an active cell of the environment with an unwanted cell.

The shaded two cells in the neighborhood of the first active cell signal the first active cell that they cannot switch to an active state. Therefore, there are six options for transmitting an active signal from eight directions. Even if the LTF points to one of the shaded neighborhood cells, the first active cell transmits an active signal in the opposite direction. In this case, the LTF result is additionally influenced by the cells in the neighborhood of the unwanted active cell. In Figure 13.4, the LTF result of the first active cell is affected by the first three cells in the neighborhood (highlighted in blue). The code of these cells indicates one of the shaded cells of the neighborhood, therefore, at the next time step, the first active cell goes to the opposite cell of the neighborhood. The second active cell transmits an active signal in accordance with its own LTF.

In the ACA environment, there may be situations where one cell is undesirable for other active ACA cells. Moreover, for this active cell there is no restriction on unwanted active cells. Therefore, this active cell freely moves across the ACA field. Thus, in the ACA environment there are active cells that "know" which active cells they should not meet, and there are also cells that "do not know" which active cells they are not allowed to meet with (there are no prohibitions).

If unwanted cells are determined by an active state, then the implementation of the ACA paradigm is quite simple. The structure of each ACA cell is slightly complicated. The number of connections between cells increases. For example, if we use the von Neumann neighborhood, and each active state is determined by the code of two cells of the neighborhood, then the number of active states is six (if we do not take into account the reverse encoding). In this case, each cell can contain six information outputs or three outputs, which represent the code of the corresponding state with a binary code (two states in such coding are redundant). Also, various active states of cells can be represented by a certain color, which significantly complicates the structure of the cell, the complexity of which is determined by the number of colors used. There are other options for the presentation of active cells.

As a result of such coding, an unwanted active cell transmits an information signal about its active state to all cells in the neighborhood. This signal appears as informational as additional information outputs of cells in the neighborhood of an unwanted active cell. Such information signals are present at additional information outputs of all cells in the neighborhood of each active cell. If at the current time step, the cell is not active and is not a cell in the neighborhood of the active cell, then the additional information outputs do not have signals of logical "1".

Since active cells themselves "do not know" that they are undesirable for other active cells, it is necessary that active cells be tuned in to signals from unwanted active cells. To solve this problem, an additional combinational circuit and additional inputs further complicate the structures of ACA cells. With the help of additional circuits and inputs, a signal is generated to

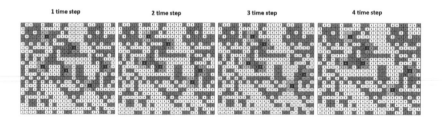

| 1 time step | 2 time step | 3 time step | 4 time step |

FIGURE 13.5
An example of the functioning of ACA and active cells in the presence of several unwanted active cells in it.

prohibit the transmission of an active signal to a given cell of the neighborhood from an active cell.

When all such settings are set, the ACA can operate in the desired modes. An example of ACA functioning with unwanted active cells is shown in Figure 13.5.

The example (Figure 13.5) shows the movement of active cells without interfering with unwanted active cells. All conditions are met and unwanted cells do not intersect. Such an ACA can function for a rather long time without long cycles of individual cells and without the formation of a large number of new cells.

Cells with blue neighborhood cells are unwanted. They move independently according to a code indicating an active state. Active cells with a yellow neighborhood bend around or move away from unwanted active cells.

The states of unwanted active cells can be established in advance at the initial moment of time or at one of the intermediate time steps. Also, an undesirable active state can be formed as a result of the LSF performance of one of the active cells, which sets the prohibition signal at one of the time steps of functioning. Similarly, as a result of performing LSF, one active cell can remove the signal that prohibits the transmission of an active signal at one of the time steps.

These models and paradigms describe the behavior of active cells independent of each other. All active cells function on the basis of the active states inherent in them and, in accordance with the LTF, choose the direction of transmission of the active signal at each time step. However, the processes of formation of colonies of active cells, as well as their interaction with each other, are little described. In Bilan, Bilan, and Motornyuk (2020), the final chapter briefly describes the formation of colonies, as well as their behavior.

A colony of active cells is a group of active cells in which each active cell can transmit a signal to any other active cell of the group, and this signal will be transmitted only through the cells belonging to the cells of the colony. All cells in the colony have the same active state and can perform as the same LSF. Before joining into a colony, active cells move according to their

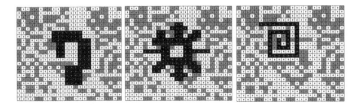

FIGURE 13.6
Examples of active cell colonies.

own LTF and during the approach of cells with the same active states, they combine and form a colony. The unification is carried out according to the properties of the active state, but not according to equal information states. An example of formed colonies of active cells is shown in Figure 13.6.

All cells in the colonies have different informational states, but they performed the same LTF prior to colony association. After combining active cells into a colony, the active cells of the colony select the main cell (leader), which further controls the movement of the entire colony. Active cells that cannot transmit a signal to each other through homogeneous cells belong to different colonies with homogeneous properties.

Colony cells contain two types of active cells:

- active cells, in the neighborhood of which there are only active cells (internal cells of the colony);
- active cells, in the neighborhood of which there are, in addition to active cells, and inactive cells (extreme cells of the colony).

The inner cells of the colony do not decide on the direction of movement of the colony, since they receive signals from the main cell, and also transmit signals from cell to cell.

All cells of the colony can move simultaneously (in one time step at a distance of one cell), or can move sequentially one after the other. The sequential movement model describes the movement of a human crowd, flocks of animals, and other group movement of automated vehicles.

Since the simultaneous movement of groups of people and animals is extremely rare in a smart city environment, the chapter discusses the forward movement of colonies. Simultaneous movement is often found in military service, where there is a need for coordinated simultaneous work by members of the entire military group. An example of the progressive movement of a group of people is shown in Figure 13.7.

In this case, the shift to the right begins from the rightmost cells of the colony. These cells, at the initial moment, "detach" from all cells of the colony at a distance of one cell and release the previous cells from the active state to

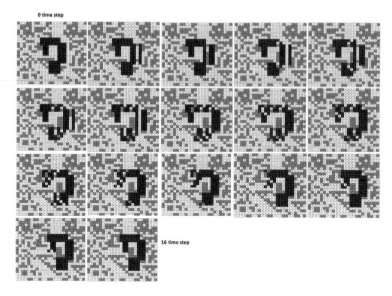

FIGURE 13.7
An example of the translational movement of all active cells of the colony to the right, taking into account the states of all neighboring active cells of the colony.

fill them with the second active cells of the colony from the right edge. The process is repeated for all other active cells in the colony. The cells transmit an active signal to the right neighboring cells to set them up in an active state. To achieve uncomplicated self-organization of active cells in the colony, no external control signal is used. To simulate real conditions, it is necessary to choose goals for the achievement of which, by means of a collective decision, the cells choose the direction of movement of the entire colony of active cells. This implementation makes the CA asynchronous. If several different colonies are present in the ACA field (formed by active cells with different active states), then sooner or later the colonies begin to interact.

13.3 Colony Formation

There are various options for the formation of colonies. One of the options is the initial formation of a colony of active cells when setting the initial conditions. This is due to the original modeling task. However, if the process is random, which depends on the initial conditions of the ACA, colonies can be created during evolution at different time steps. If you do not set various attitudes and restrictions, then chaotically formed colonies are unstable and quickly disintegrate. They combine active cells with different active states. Such colonies are formed from newly formed cells by the active cells of the colony themselves. In these colonies, active cells constantly disappear and new ones appear.

This chapter describes resistant colony forms that are formed by active cells with the same active states. When cells with the same active states meet, new active cells are not formed and already existing interacting active cells are not destroyed.

For the formation of a colony, it is necessary to set the initial conditions within which the colony is formed. Initially, active cells with different active states move in the ACA field. Also new active cells are formed and old active cells disappear. The formation of a new active cell is carried out by the interaction of active cells with different active states. If two active cells with the same active states become adjacent (they are neighborhood cells for one another), then they combine and form a colony. Subsequently, these active cells move together within the directions of the colony movement. The colony formation process is shown in Figure 13.8.

Once active cells are adjacent, they form the beginning of a colony. An important aspect is the selection of the main active cell of the colony. The simplest way to select a chief cell is to analyze the states of active cells forming a colony. The main one is the active cell, in the neighborhood of which there are more cells with a state of logical "1". The own state of each active cell can also be taken into account. In the next time steps, the cells of the colony move in the directions indicated by the main cell of the colony.

This situation is described for the case when there are only two active cells. However, colonies with the same active states of cells can be created independently and can have the same active states. Each of these colonies has its own main (chief) active cell in the colony. Formed homogeneous colonies can meet and merge into one large colony. When these colonies merge, it

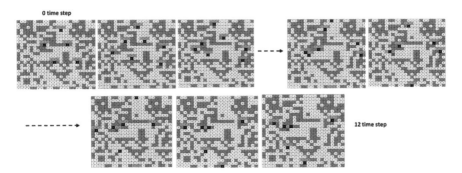

FIGURE 13.8
An example of the formation of a colony of active cells with the same states.

becomes necessary to select the main active cell of the new colony from several main active cells.

When these colonies merge, it becomes necessary to select the main active cell of the new colony from several main active cells:

- By the largest number of cells that form a colony;
- By the largest number of cells in the state of logical "1" and belonging to the neighborhood of the main cell of one of the colonies;
- selection of the main cell of the newly formed colony, which is not one of the previous main cell forming colonies.

Examples of colony merger for each of the options are shown in Figure 13.9.

Each newly formed colony has only one main cell. This cell controls the further movement of the colonies. It is this cell that decides which direction to move. In the cell with the first mode of the third variant, the cells through which the active signal passed are highlighted in yellow. A new main colony cell is formed in the cell where two signals meet. If there are several such cells, then one of the cells is selected, which has more cells in the neighborhood that have a state of the logical "1". If there are equal numbers of such cells in the neighborhood of both selected cells, then the colonies do not merge in order to avoid conflict situations. They move across the ACA field and change. During the next meeting of the colonies, one of the colonies can already

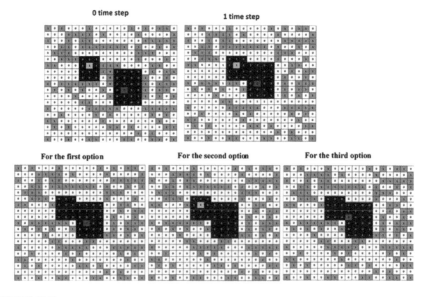

FIGURE 13.9
Examples of colony fusion.

dominate the other (stronger than the other), which simplifies the choice of a new main cell of the new united colony. The association and formation of colonies, as well as the choice of the main cell of the colony, is carried out within the framework of the permissible functioning of ACA.

13.4 Cell Colony Movement

When colonies form, they constantly move across the ACA field. All biological associations are also moving. Colonies can only stop if the environment is favorable to keep the colony alive. Depending on the environment, the colonies move toward a favorable environment. Movement and direction of movement can be caused by various factors. The environment is assessed by the outermost cells of the colony. Signals from the outermost cells through the inner cells are transmitted to the main cell of the colony and, based on the information received, the main cell indicates the direction of movement by transmitting a control signal to the corresponding outermost cells of the colony, which begin to move.

If the direction of movement is chosen only by the extreme cells, then different directions of movement can be chosen, which will lead to the stretching and rupture of the colony into different parts. An example of such a gap is shown in Figure 13.10.

In Figure 13.10, different directions of movement are represented by different colors. As a result of this situation, middle cells of the colony may remain, which cannot choose the desired direction of movement. Therefore, the choice can be made according to priorities. Such priorities can be the following:

- priority according to the number of active cells of the neighborhood with the selected direction of shift;
- by the number of active cells in the neighborhood that are in a state of logical "1" (or logical "0") and have the same direction of movement;

FIGURE 13.10
An example of a colony rupture when the direction of movement is chosen by the edge cells of the colony.

Both priorities do not provide a complete solution to the problem of choosing the direction of travel. This mode is especially unacceptable when the movement must be unidirectional. In this case, there are two options for choosing the direction of movement:

- control mode based on "voting" and the strength of the outermost cells of the colony;
- all extreme cells of one direction make the same decision and there are more of these cells than cells of the other direction;
- movement toward the extreme border inactive cell, which has a state of logical "1" and is an extreme single cell when going around the boundaries of the colony clockwise. The beginning of the walk can be accepted in an arbitrary extreme cell of the colony.

The first mode is based on comparing opposite extreme cells on the same line (one line of the image matrix). What extreme cells are stronger other edge cells and this cells are located on one edge and them more on this edge of the colony, in that direction and the movement is carried out. However, such a mode is difficult in its implementation in the paradigms of classical CA. However, it is more democratic.

The second mode is close to the first mode and is often more effective. However, situations may arise when in several directions the number of edge cells is the same and they all made the same decisions regarding the direction of movement, then in this case the colony does not move in any direction.

The most effective mode is the mode of control of movement of the main cell of the colony. There are also several options for controlling the movement of the colony using the main cell of the colony.

- choosing of direction by movement using the analysis of the states of active cells of the neighborhood (movement blindly);
- choosing the direction of movement by analyzing the states of the edge cells of the neighborhood;
- selection of the direction of movement using the analysis of inactive cells of the environment bordering on the edge cells of the colonies.

The first mode of choosing the direction of movement of the colony is the simplest to implement. However, this regime can lead the entire colony to an undesirable place (in an unfavorable environment) for the entire colony. An example of the movement of a colony in accordance with the first regime is shown in Figure 13.11.

The signal of the direction of movement is transmitted to the cells of the colony and does not reach all the cells of the colony at the same time. There are cells in the colony that have not yet received a directional control signal.

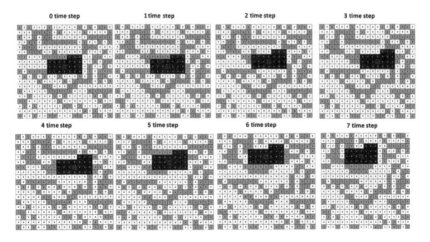

FIGURE 13.11
An example of choosing the direction of movement of a colony based on the main cell of the colony and the first control mode.

Therefore, the movement of the entire colony is carried out with small gaps and in the process of movement the colony is pulled in the direction of movement. After moving all active cells of the colony, the shape of the colony takes on the previous geometric shape. In this case, one time step of the colony movement per one ACA cell is stretched by the number of aperiodic steps, which is determined by the shape of the colony and the number of active cells that make up the colony.

Figure 13.11 does not show the process of stretching the colony in one time step of movement. These steps are skipped. The final state is shown after moving one time step of all active cells of the colony. The stretching process in is shown in Figure 13.7.

The second mode is implemented so that the main cell of the colony expects signals from the edge cells of the colony. The edge cells of the colony analyze neighboring inactive cells and determine the number of neighboring states of logical "1" and "0". If the number of logical "1" exceeds the number of logical "0", then the outer cell of the colony forms an information signal, which is transmitted through the cells of the colony to the main cell of the colony. The main cell of the colony, having received the information signal, determines the direction from which the signal came and generates a control signal that the direction of movement is carried out in the direction of the appearance of the information signal. If several edge cells simultaneously form an information signal, then the main cell chooses the direction from which the information signal first came. Figure 13.12 shows an example of choosing the direction of movement of the colony in the second mode.

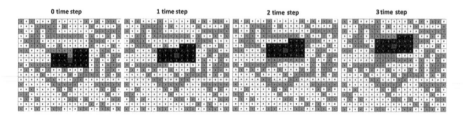

FIGURE 13.12
An example of choosing the direction of movement of a colony by the main active cell according to the second mode.

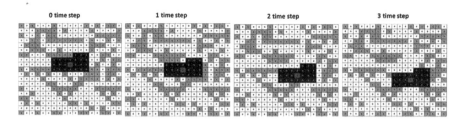

FIGURE 13.13
An example of choosing the direction of the main active cell by sectors consisting of a large number of ones cells.

The second mode is more targeted for the movement of the colony and also requires more time.

The yellow cell at the zero time step has a larger number of cells with a ones state, and it is also located closer to other such extreme cells of the colony. Therefore, the active signal of the direction of movement arrives at the main cell before other cells. The entire colony moves in the direction of this cell for a given number of steps. If you use this analysis after each time step, then the colony will move in different directions after each time step. Therefore, the target option is selected for several time steps of movement toward the goal.

The third mode provides for the choice of the direction of movement by the simultaneous analysis of all border inactive cells with the cells of the colony. Based on this analysis, the sector is determined with the largest number of border inactive cells that have a state of logical "1". The movement is carried out in the direction of this sector. Figure 10.13 shows an example of choosing the direction of movement by the main active cell in the third mode.

At the bottom right of the figure, a 3 × 3 field is highlighted, which contains the largest number of cells with a state of logical "1" among all such fields along all the outer cells of the colony. Therefore, the main cell of the colony chooses this direction, and the entire colony moves in this direction. As can be seen from Figure 13.13, in the selected direction, the highest concentration of cells with a state of logical "1". This corresponds to the purpose of the colony.

Colony travel time depends on the geometric shape of the colony and on the direction of movement. The more active cells located on one straight line of movement, the slower the movement of the entire colony.

13.5 Interaction of Active Cell Colonies

Colony unification is the interaction of small colonies of different geometric shapes and the formation of new large colonies with specific geometric shapes. At the same time, nothing is said about the fact that colonies can become smaller and geometric shapes also change. In addition, colonies can disintegrate into smaller.

What can affect the reduction or destruction of the colony? Colony breakdown may depend on the time it takes for the control signal to reach the colony cell. If the transmission time of the control signal exceeds a predetermined time, the cell colony becomes less mobile, which leads to long delays in its movement. This colony is not optimal for long-term survival. Therefore, the removed cells are rejected, or they themselves detach from the colony and "choose" a new main cell. It can also happen with small colonies that are elongated in shape with the main call at the edge of the colony. Examples of such colonies are shown Figure 13.14.

How do such colonies disintegrate? Colony cells, to which, after a given number of time steps, the control signal does not reach the remote cells of the colony, then these cells begin to move independently and, when interacting with other cells, can form another colony with cells of the same properties. For new, formed colonies, the task of choosing the main cell of the colonies arises. The choices for new colony main cells have been described previously.

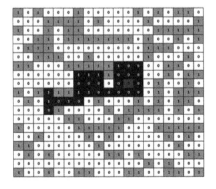 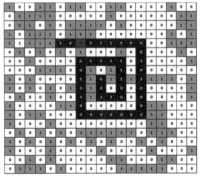

FIGURE 13.14
Examples of geometric shapes of cell colonies that can divide into multiple colonies.

This collapse of colonies is called self-disintegration, which is carried out due to internal contradictions.

However, colonies can disintegrate or change in size and geometric shape under various external influences. Such influences may be present in colonies with other active properties or unwanted cells.

If there are several colonies of active cells with different active states, then a conflict may arise in the struggle for the environment, which leads to the destruction of one colony by another. If both colonies move in the same direction of a favorable habitat from different points of location, then there is a possibility of their meeting and interaction. Since the colonies have different active cells, the conflict is inevitable. As soon as the outer cells of the colonies become adjacent, they begin to destroy each other. Stronger cells in one colony destroy weaker cells in another colony.

The strength of each active cell of colonies is determined by the state of its neighborhood of cells. Let's present several options for determining strong cells:

- the strength of a cell is determined by the number of cells in the neighborhood that have a logical state of "1", as well as their own state;
- the strength of the cells is determined by the binary code of the neighborhood cells, which determine its active state;
- the strength of a cell is determined by the number of neighboring cells in the neighborhood that have the same active state, as well as have a logical state "1".

When the border cells of several colonies meet, these cells begin to analyze the states of the cells of the neighborhood and their own states. For the first variant of interaction by each border (edge) cell of the colony, which interacts with the border cells of another colony, the analysis of the states of the neighborhood cells is performed and the number of single states of the cells of the neighborhood is determined. The edge cells of one colony, which have a large number of single cells in the neighborhood, survive, and weak neighboring cells of another colony, which have fewer ones cells of the neighborhood, "die" and the colony size decreases by one cell. Weak cells may not "die", but become a cell of a stronger colony (captured) and take on the active state of the cells that have captivated it. Figure 13.15 shows an example of the interaction of two colonies with different active cells according to the first variant of strong cells.

Figure 13.15 shows two colonies with different properties. The right colony moves up the cell field, and the left colony moves to the left toward the right colony. Upon meeting, weak cells according to the first option die (highlighted in blue), since they have fewer cells in the vicinity that have a state of the logical "1". In this example, all border cells of the right colony

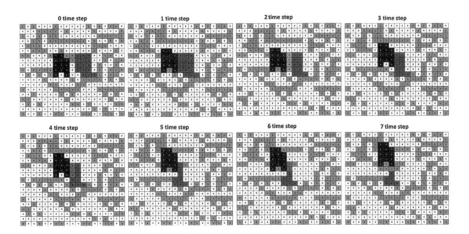

FIGURE 13.15
An example of the interaction of two colonies with different active cells according to the first variant of strong cells.

turned out to be weak. In the last step, the two colonies diverge; however, the right colony has decreased as a result of collision and interaction with the left colony of cells.

As you can see from the example, colonies can move in different directions at each time step and can interact with each other while moving. However, it is obvious that for the first variant the cells of both colonies can have the same strength and at this moment the cells are not destroyed and continue to move along with all the cells of the colony. And if the colonies continue to move toward each other, then the movement of cells of equal strength does not occur, since the cell can only move to a free cell (not an active cell). To implement this mode, the edge cells of the colonies can interact only with the cells of another colony lying on the same line of direction of movement.

The second option for determining the strength of a cell is similar to the first option. Moreover, the number of single cells in the neighborhood does not affect the strength of the cell. An example of the interaction of two colonies according to the second variant of determining the strength of the cell is shown in Figure 13.16.

In this case, other cells can be destroyed and those that were strong for the first option can become weak. Colonies also take on other geometric shapes.

The cells of the left colony determine their strength by the top three cells of the Moore neighborhood, and the cells of the right colony determine their strength by the left three cells of the Moore neighborhood.

In this case (Figure 13.16), cells of the left colony are already destroyed (which did not happen in the example of the previous variant), since some of the edge cells of the right colony are "stronger" than the edge cells of the left colony. At the sixth time step, two cells from different colonies claim one

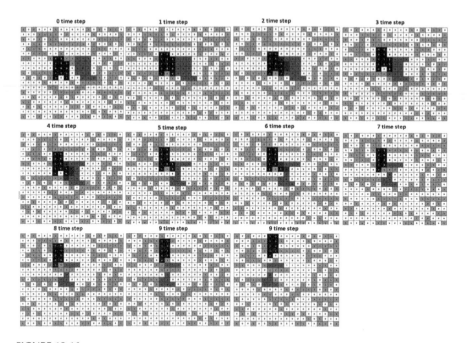

FIGURE 13.16
An example of the interaction of colonies with different active cells according to the second variant of determining cell strength.

space cell, which is freed by the dead cell of the left colony. In this case, the cell is occupied by the cell of the left colony, since its code corresponds to 7 (111) in the binary system, and the cell code of the right colony is 2 (010), which means the weakness of the right cell. Some interacting cells are not destroyed because they have equal forces, and also some cells do not move because there are no free space cells in the path of its movement.

The third option for determining the strength of the cell can be implemented independently, or it can be used in conjunction with the first or second option for determining the strength of the cell. It is very convenient to use in a situation where there are several equivalent cells from each colony. In this situation, the cell of the colony dies, in the neighborhood of which there are more cells from the opposite colony. An example of the interaction of colonies according to the third option for determining the strength of cells is shown in Figure 13.17.

Figure 13.17 shows that the cells of both colonies are destroyed, depending on the state of the colony cells belonging to each colony. During movement, weak cells of the colonies are also destroyed if they interact.

For all variants of interaction of colony cells, the interaction of those cells that have common sides is used. If there is a diagonal contact of the cells of the colony (contact of the cells that place in the vertex of colony), then such

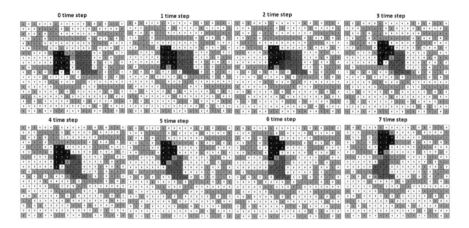

FIGURE 13.17
An example of the interaction of colonies with different active cells according to the third variant of determining cell strength.

interaction is ignored. However, the developer can take this interaction into account as well. There are many options for colony interactions. The choice of the option depends on the dynamic process that is modeled and on the task.

Such modeling allows to describe the game life on a new level and in new paradigms.

The collective movement models of humans and animals described in this chapter can be useful for designing and building smart urban environments. If we divide the city into sectors and describe them with a model in the environment of CA, we can predict the further behavior of groups of people and flocks of animals in the city. Also, in the event that people or animals appear in the city who have become ill with severe viral diseases, it is possible to predict their movement and avoid unwanted meetings with them. In addition, such a model is imperfect and can be changed and improved for different situations.

References

Abidi, B., and M.A. Abidi. 2007. *Face Biometrics for Personal Identification: Multi-Sensory Multi-Modal Systems* (Signals and Communication Technology). Springer.

Afolabi, A., and D. Ademiluyi. 2015. Ear Symmetry Determination in Ear Based Human Recognition. *Institute of Ear Biometric Research*, 5(6). http://www.iebre-search.org/1194765-88.pdf.

Barlow, M., and C. Levy-Bencheton. 2018. *Smart Cities, Smart Future: Showcasing Tomorrow*. Wiley.

Bedrouni, A., R. Mittu, A. Boukhtouta, and J. Berger. 2009. *Distributed Intelligent System*. Springer US.

Belan, S.N., and R.L. Motornyuk. 2013. Extraction of Characteristic Features of Images with the Help of the Radon Transform and Its Hardware Implementation in Terms of Cellular Automata. *Cybernetics and Systems Analysis*, 49(1), 7–14.

Belan, S., and S. Yuzhakov. 2013a. A Homogenous Parameter Set for Image Recognition Based on Area. *Computer and Information Science, Published by Canadian Center of Science and Education*, 6(2), 93–102.

Belan, S., and S. Yuzhakov. 2013b. Machine Vision System Based on the Parallel Shift Technology and Multiple Image Analysis. *Computer and Information Science, Published by Canadian Center of Science and Education*, 6(4), 115–124.

Bhuptani, M., and S. Moradpur. 2007. *RFID Technologies at the Service of Your Business = RFID Field Guide: Deploying Radio Frequency Identification Systems*, ed. Troitsky, N. Moscow: Alpina Publisher, 70–290.

Bilan, S. 2014. Models and Hardware Implementation of Methods of Pre-processing Images Based on the Cellular Automata. *Advances in Image and Video Processing*, 2(5), 76–90.

Bilan, M. 2019. Image Processing and Pattern Recognition Based on Artificial Models of the Structure and Function of the Retina. In *Handbook of Research on Intelligent Data Processing and Information Security Systems*, ed. S.M. Bilan and S.I. Al-Zoubi, 360–373. IGI Global.

Bilan, S. 2017. *Formation Methods, Models, and Hardware Implementation of Pseudorandom Number Generators: Emerging Research and Opportunities*. IGI Global.

Bilan, S., M. Bilan, and A. Bilan. 2021a. Interactive Biometric Identifcation System Based on the Keystroke Dynamic. In *Biometric Identification Technologies Based on Modern Data Mining Methods*, ed. S.M. Bilan and S.I. Al-Zoubi, 39–58. Springer.

Bilan, M., A. Bilan, and S. Bilan. 2021b. Identification of a Person by Palm Based on an Analysis of the Areas of the Inner Surface. In *Biometric Identification Technologies Based on Modern Data Mining Methods*, ed. S. Bilan, M. Elhoseny, and D.J. Hemanth, 135–146. Springer.

Bilan, S., M. Elhoseny, and D.J. Hemanth. 2021c. *Biometric Identification Technologies Based on Modern Data Mining Methods*. Springer.

Bilan, S.M., M.M. Bilan, and R.L. Motornyuk. 2020. *New Methods and Paradigms for Modeling Dynamic Processes Based on Cellular Automata*. IGI Global.

Bilan, S., R. Motornyuk, S. Bilan, and O. Galan. 2021d. User Identifcation Using Images of the Handwritten Characters Based on Cellular Automata and Radon Transform. In *Biometric Identification Technologies Based on Modern Data Mining Methods*, ed. S. Bilan, M. Elhoseny, and D.J. Hemanth, 91–103. Springer.

Bilan, S., R. Motornyuk, and S. Bilan 2014a. Method of Hardware Selection of Characteristic Features Based on Radon Transformation and Not Sensitive to Rotation, Shifting and Scale of the Input Images. *Advances in Image and Video Processing*, 2(4), 12–23.

Bilan, S., R. Motornyuk, S. Yuzhakov, and M. Halushko. 2021e. Gait Identifcation Based on Parallel Shift Technology. In *Biometric Identification Technologies Based on Modern Data Mining Methods*, ed. S. Bilan, M. Elhoseny, and D.J. Hemanth, 75–89. Springer.

Bilan, S., S. Yuzhakov, and S. Bilan. 2014b. Saving of Etalons in Image Processing Systems Based on the Parallel Shift Technology. *Advances in Image and Video Processing*, 2(6), 36–41.

Bilan, S., and S. Yuzhakov. 2018. *Image Processing and Pattern Recognition Based on Parallel Shift Technology*. CRC Press, Taylor & Francis Group.

Boissier, O., R.H. Bordini, J. Hubner, and A. Ricci. 2020. *Multi-Agent Oriented Programming: Programming Multi-Agent Systems Using JaCaMo* (Intelligent Robotics and Autonomous Agents series). The MIT Press.

Boulgouris, N.V., K.N. Plataniotis, and E. Micheli-Tzanakou. 2009. *Biometrics: Theory, Methods, and Applications*, vol. 9. Wiley-IEEE Press.

Bourlai, T. 2019. *Face Recognition Across the Imaging Spectrum*. Springer.

Buimistryuk, G. 2014. Technologies of Fusion of Sensory Information for Control in Critical Situations. *Control Engeneering Russia*, 5(53), S47–S51.

Cardullo, C., C. Di Feliciantonio, and R. Kitchin. 2019. *The Right to the Smart City*. Emerald Publishing Ltd.

Deans, S.R., and S. Roderick. 1983. *The Radon Transform and Some of Its Applications*. Wiley.

Dorogov, A. Yu. 2011. *Theoretical Foundations of Learning Algorithms for Fast Transformations*. In *Abstracts of the Conference "Technical Vision in Control Systems – 2011"*, 107–109.

Dudgeon, D.E., and R.M. Mersereau. 1984. *Multidimensional Digital Signal Processing*. Prentice Hall.

Esbensen, K. 2003. *Analysis of Multivariate Data*. Barnaul: Publishing House Alt. University.

Ell, T.A., N.L. Bihan, and S.J. Sangwine. 2014. *Quaternion Fourier Transforms for Signal and Image Processing* (Focus series), 1st ed., Wiley-ISTE.

Fairhurst, F. 2018. *Biometrics: A Very Short Introduction*. Oxford University Press.

Fairhurst, M. 2019. *Biometrics: A Very Short Introduction* (Very Short Introductions). Oxford University Press.

Ferrari, G. 2010. *Sensor Networks: Where Theory Meets Practice* (Signals and Communication Technology). Springer.

Ghaeminia, M.H., S.B. Shokouhi, and A. Amirkhani. 2021. Biometric Gait Identifcation Systems: From Spatio-Temporal Filtering to Local Patch-Based Techniques. In *Biometric Identification Technologies Based on Modern Data Mining Methods*, ed. S. Bilan, M. Elhoseny, and D.J. Hemanth, 19–37. Springer.

Golay, M.J.E. 1969. Hexagonal Parallel Pattern Transformations. *IEEE Trans. Computers*, vol. C-18, 733–740.

Gonzalez, R.C., R.E. Woods, and S.L. Eddins. 2004. *Digital Image Processing Using MATLAB*. Pearson Education.

Gonzalez, R.C., and R.E. Woods. 2008. *Digital Image Processing*, 3rd ed. Prentice Hall.

Hassanien, A.E., M. Elhoseny, S.H. Ahmed, and A.K. Singh. 2019. *Security in Smart Cities: Models, Applications, and Challenges*. Springer International Publishing.

Holland, J. 1966. Universal Spaces: A Basis for Studies in Adaptation. In *Automata Theory*, ed. E.R. Caianiello, 218–230. Academic Press.

Iannacci, J. 2017. *Introduction to MEMS and RF-MEMS: From the Early Days of Microsystems to Modern RF-MEMS Passives*. IOP Publishing Ltd.

Jain, A. 1989. *Fundamentals of Digital Image Processing*. Prentice Hall, Chap. 9.

Karaboga, D.D. 2005. *An Ider Based on Honey Be Swarm for Numerical Optimization, Technical Report-TR06*. Erciyes University, Engineering Faculty, Computer Engineering Department.

Kazantseva, A.G. 2013. Identification of a Person by Gait with the use of Wearable Sensors. Overview Research. *Mathematical Structures and Modeling*, 2(28), 103–111.

Klems, M. 2005. *RFID: Transport und Logistik an der Schwelle eines neuen Zeitalters?* (German Edition). GBI-Genios Verlag.

Krichevsky, R.L., and E.M. Dubovskaya. 2009. *Social Psychology Small Group*. Aspect Press.

Krit, S., V.E. Bălaş, and M. Elhoseny. 2020. *Sensor Network Methodologies for Smart Applications*. Information Science Reference.

Kulkarni, A.J., K. Tai, and A. Abraham. 2015. *Probability Collectives: A Distributed Multi-Agent System Approach for Optimization* (Intelligent Systems Reference Library Book 86). Springer.

Savoy, S., J.J. Lavigne, M.B. Clevenger, J.E. Ritchie, B. McDoniel, S. Yoo, E.V. Anslyn, J.T. McDevitt, J.B. Shear, and D. Neikirk. 1998. *Solution-Based Analysis of Multiple Analytes by a Sensor Array: Toward the Development of an Electronic Tongue. In SPIE Conference on Chemical Microsensors and Applications*, SPIE Vol. 3539, Boston, MA, November 4.

Lesser, V., and Charles L. Ortiz, Jr. 2003. *Milind Tambe: Distributed Sensor Networks*. Springer US.

Mersereau, R.M. 1979. The Processing of Hexagonally Sampled Two Dimensional Signals. *Proc. IEEE*, 67(6), 930–949.

Liu, Z. 2010. *Investigations on Multi-Sensor Image System and Its Surveillance Applications*. Dissertation.Com.

Minichino, J., and J. Howse. 2015. *Learning OpenCV 3 Computer Vision with Python*. Packt Publishing – ebooks Account.

Motornyuk, R., A. Bilan, and S. Bilan. 2021. Research of Biometric Characteristics of the Shape of the Ears Based on Multi-Coordinate Methods. In *Biometric Identification Technologies Based on Modern Data Mining Methods*, ed. S. Bilan, M. Elhoseny, and D.J. Hemanth, 177–194. Springer.

Motornyuk, R.L., and S. Bilan. 2019a. Methods for Extracting the Skeleton of an Image Based on Cellular Automata with a Hexagonal Coating Form and Radon Transform. In *Handbook of Research on Intelligent Data Processing and Information Security Systems*, ed. S.M. Bilan and S.I. Al-Zoubi, 289–329. IGI Global.

Motornyuk, R.L., and S. Bilan. 2019b. The Moving Object Detection and Research Effects of Noise on Images Based on Cellular Automata with a Hexagonal Coating Form and Radon Transform. In *Handbook of Research on Intelligent Data Processing and Information Security Systems*, ed. S.M. Bilan and S.I. Al-Zoubi, 330–359. IGI Global.

Müller, H.J., and R. Dieng. 2012. *Computational Conflicts: Conflict Modeling for Distributed Intelligent Systems*. Springer.

Nejati, H., L. Zhang, T. Sim, E. Martinez-Marroquin, and G. Dong. 2012. Wonder Ears: Identifcation of Identical Twins from Ear Images. In *Proceedings of the International Conference on Pattern Recognition*, 1201–1204. New York: IEEE.

Nixon, M.S., and A.S. Aguardo. 2002. *Feature Extraction and Image Processing*. Newnes.

Patel, H.K. 2013. *The Electronic Nose: Artificial Olfaction Technology*. Springer.

Peng, S., M.N. Favorskaya, and H. Chao. 2020. *Sensor Networks and Signal Processing*. In *Proceedings of the 2nd Sensor Networks and Signal Processing (SNSP 2019)*, 19–22 November 2019, Hualien, Taiwan, p. 176. Springer.

Picon, A. 2015. *Smart Cities: A Spatialised Intelligence Paperback*. Wiley.

Pratt, W.K. 2016. *Digital Images Processing*, 3rd ed. Wiley.

Richard, O., and P.E. Duda. 1972. Hart Use of the Hough Transformation to Detect Lines and Curves in Pictures. *Published in the Comm. ACM*, 15(1), 11–15.

Riul, A., Jr., C.A.R. Dantas, C.M. Miyazaki, and O.N. Oliveira. 2010. Recent Advances in Electronic Tongues. *The Analyst*, 135(10), 2481. https://doi.org/10.1039/c0an00292e.

Ryan, Ø. 2019. *Linear Algebra, Signal Processing, and Wavelets – A Unified Approach: Python Version* (Springer Undergraduate Texts in Mathematics and Technology), 1st ed. Springer.

Ryzko, D. 2020. *Modern Big Data Architectures: A Multi-Agent Systems Perspective*. Wiley.

Saračević, M., M. Elhoseny, A. Selimi, and Z. Lončeravič. 2021. Possibilities of Applying the Triangulation Method in the Biometric Identifcation Process. In *Biometric Identification Technologies Based on Modern Data Mining Methods*, ed. S. Bilan, M. Elhoseny, and D.J. Hemanth, 1–17. Springer.

Solomon, C., and T. Breckon. 2011. *Fundamental of Digital Image Processing: A Practical Approach with Examples in Matlab*. Wiley-Blackwell.

Song, H., R. Srinivasan, T. Sookoor, and S. Jeschke. 2017. *Smart Cities: Foundations, Principles, and Applications*. Wiley.

Staunton, R.C. 1989a. *Hexagonal Image Sampling: A Practical Proposition*. In *Proc. SPIE*, vol. 1008, 23–27.

Staunton, R.C. 1989b. The Design of Hexagonal Sampling Structures for Image Digitisation and Their Use with Local Operators. *Image and Vision Computing*, 7(3), 162–166.

Sun, E.C. 2011. *Ant Colonies: Behavior in Insects and Computer Applications* (Computer Science, Technology and Applications). Nova Science Pub Inc.

Symond, J., J. Ayoade, and D. Parry. 2009. *Auto-Identification and Ubiquitous Computing Applications*. IGI Global.

Tinder, R.F. 2007. *Relativistic Flight Mechanics and Space Travel: A Primer for Students, Engineers and Scientists*. Morgan and Claypool Publishers.

Toko, K. 2003. Integrated Analytical Systems. In *Comprehensive Analytical Chemistry*, ed. S. Alegret, vol. XXXIX. Amsterdam: Elsevier.

Tou, J.T., and R.C. Gonzalez. 1974. *Pattern Recognition Principles*. Addison-Wesley.

Yamagata, Y. 2020. *Urban Systems Design: Creating Sustainable Smart Cities in the Internet of Things Era*. Elsevier.

Yuzhakov, S. 2019. Reproduction of Images of Convex Figures by a Set of Stored Reference Surfaces. In *Handbook of Research on Intelligent Data Processing and Information Security Systems*, ed. S.M. Bilan and S.I. Al-Zoubi, 264–288. IGI Global.

Yin, S., and F.T.S. Yu. 2002. *Fiber Optic Sensors*. CRC Press.

Yuzhakov, S., S. Bilan, S. Bilan, and M. Bilan. 2021. Search of Informative Biometric Characteristic Features of the Palm Based on Parallel Shift Technology. In *Biometric Identification Technologies Based on Modern Data Mining Methods*, ed. S. Bilan, M. Elhoseny, and D.J. Hemanth, 147–158. Springer.

Yuzhakov, S., and S. Bilan. 2019. Identification System for Moving Objects Based on Parallel Shift Technology. In *Handbook of Research on Intelligent Data Processing and Information Security Systems*, ed. S.M. Bilan and S.I. Al-Zoubi, 374–387. IGI Global.

Woldridge, M.J., and N.R. Jennings. 1994. *Agent Theories, Architectures and Languages: A Survey – Intelligent Agents. In ECAI-94: Workshop on Agent Theories, Architectures and Languages*, Amsterdam, August 8–9, pp. 3–39.

Woodward, M. 1984. *Francis Muir Hexagonal Sampling*. Stanford Exploration Project, SEP-38, 183–194.

Zalmanzon, L.A. 1989. *Fourier Transform, Walsh, Haar and Their Application in Control, Communications and Other Areas*. Moscow: Nauka.

Zheng, S., J. Zhang, K. Huang, R. He, and T. Tan. 2011. *Robust View Transformation Model for Gait Recognition*. In *Proceedings of the IEEE International Conference on Image Processing*, 2073–2076.

Index

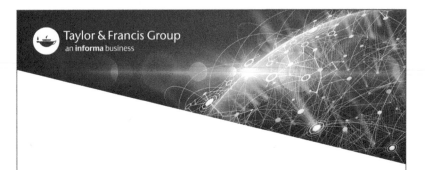